# CSR, Sustainability, Ethics & Governance

**Series Editors**

In recent years the discussion concerning the relation between business and society has made immense strides. This has in turn led to a broad academic and practical discussion on innovative management concepts, such as Corporate Social Responsibility, Corporate Governance and Sustainability Management. This series offers a comprehensive overview of the latest theoretical and empirical research and provides sound concepts for sustainable business strategies. In order to do so, it combines the insights of leading researchers and thinkers in the fields of management theory and the social sciences – and from all over the world, thus contributing to the interdisciplinary and intercultural discussion on the role of business in society. The underlying intention of this series is to help solve the world's most challenging problems by developing new management concepts that create value for business and society alike. In order to support those managers, researchers and students who are pursuing sustainable business approaches for our common future, the series offers them access to cutting-edge management approaches.

*CSR, Sustainability, Ethics & Governance* is accepted by the Norwegian Register for Scientific Journals, Series and Publishers, maintained and operated by the Norwegian Social Science Data Services (NSD)

Franzisca Weder • Lars Rademacher •
René Schmidpeter
Editors

# CSR Communication in the Media

## Media Management on Sustainability at a Global Level

Springer

*Editors*
Franzisca Weder
School of Communication and Arts
University of Queensland
Brisbane, QLD, Australia

Lars Rademacher
Department Media
Darmstadt University of Applied Sciences
Darmstadt, Germany

René Schmidpeter
M3TRIX Institute of Sustainable Business
Cologne, Germany

ISSN 2196-7075 ISSN 2196-7083 (electronic)
CSR, Sustainability, Ethics & Governance
ISBN 978-3-031-18975-3 ISBN 978-3-031-18976-0 (eBook)
https://doi.org/10.1007/978-3-031-18976-0



This Springer imprint is published by the registered company Springer Nature Switzerland AG
The registered company address is: Gewerbestrasse 11, 6330 Cham, Switzerland

# Contents

# Sustainable Communication? Media and Communication Responsibility in Global Transformation Processes

Franzisca Weder, Lars Rademacher, and René Schmidpeter

## 1 Sustainability as Guiding Principle for CSR?

Sustainability is currently one of the most challenging terms and labels, used and abused in various public communication efforts. Over the past few years, mainly corporate communication transformed the simple meaning of the ability of a system or structure to be maintained at a certain rate or level into a "label" or "buzz word," now associated with "green" or "future oriented," adding a certain degree of morality and the feeling of "you've done well."

The concept of sustainability is originally focused on how we treat our resources (WCED, 1987), it is still communicated as alternative within our market-driven economy and very rarely as alternative *to* the existing capitalistic system.

However, even within the existing economic system, not only corporates, but also organizations of all kinds and sizes are responsible toward the society, which is part of the definition of the so-called Corporate Social Responsibility (European Greenbook, 2001). Furthermore, they play a major role in socio-environmental transformation processes, as soon as they start to think and communicate about possible alternative futures. To conceptualize and frame a more sustainable future, the Sustainable Development Goals (SDGs, UN, 2022) offer a framework or

F. Weder (✉)
School of Communication and Arts, University of Queensland, Brisbane, QLD, Australia
e-mail: f.weder@uq.edu.au

L. Rademacher
Department Media, Darmstadt University of Applied Sciences, Darmstadt, Germany
e-mail: lars.rademacher@h-da.de

R. Schmidpeter
M3TRIX Institute of Sustainable Business, Cologne, Germany
e-mail: schmidpeter@m3trix.de

 
F. Weder et al. (eds.), *CSR Communication in the Media*, CSR, Sustainability, Ethics & Governance, https://doi.org/10.1007/978-3-031-18976-0_1

"blueprint" to achieve a better future for all. Thus, to define the future, sustainability works as normative and guiding principle (Weder, 2021a; Grunwald, 2012).

Sustainability as guiding principle is part of political debates, but much more it is taken up by corporates and industry as part of the responsibility they want to take toward the society—again, in an economic, but also social, environmental, and cultural dimension (Rasche et al., 2017; Diehl et al., 2017; Weder et al., 2019). Corporate Social Responsibility is predominantly directed toward specific stakeholders but includes negotiations and conversations within organizational settings about how to take the allocated responsibility and the fit of the activities related to the core business. Here communication comes into play.

## 2 CSR and Communication

CSR Communication includes all communication within the organizational context, where responsibilities in an economic, social, and environmental dimension are discussed. Thus, it is a way of thinking, related to the responsibility toward the society that an organization is willing to take. However, CSR communication puts the interests and requirements of (specific) stakeholders at the center of the organizations' attention (Freeman & Moutchnik, 2013) and focuses on communication as dialogue, on participatory approaches and engagement (Golob & Podnar, 2011). So, CSR communication has advanced from communication about certain CSR activities to a holistic, inclusive understanding of conversations about the relationship between economic and ethical values (Christensen et al., 2017).

Over the years, CSR and related communication efforts have been operationalized from a pragmatic perspective, because stakeholders request companies to not only be engaged, but also to engage the stakeholders themselves and to keep them informed and involved in related organizational matters through communication. Driven by this demand, CSR communication is first and foremost a straightforward task, focused predominantly on keeping the stakeholders informed about corporate CSR activities, with some downsides from aspirational CSR communication (Winkler et al., 2020) to greenwashing (Elving et al., 2015). However, CSR related communication is happening on various levels and includes, for example, leadership communication, community engagement, and issue and crisis management, to name a few. Communicating about and for an organization and their responsibilities toward the society in multiple dimensions implies more that only better information; it implies a deep understanding of the central role of communication, the media and strategic communication and Public Relations for social impact, stakeholder engagement and getting the social license to operate (Hurst & Johnston, 2021; Johnston & Lane, 2021). It includes increased awareness of relevant issues, the willingness to leave the own comfort zone, the willingness to change and to negotiate the meaning of core values that drive CSR activities within the organization (Schoeneborn & Trittin, 2013; Trittin & Schoeneborn, 2017), from the leadership to the individual employee.

Today, after decades of practical work in this area and an increasing amount of academic work in the area of CSR, Corporate Citizenship and Corporate Governance and related communication activities, there are three developments that we want to highlight as future directions—specifically from a media and communication perspective and considering organizational activities that are linked to global transformation processes and the principle of sustainability: (1) the necessity to re-define and reframe sustainability as normative principle and *guardrails* for corporate responsibility; (2) the necessity to conceptualize *communicative sustainability* and sustainable communication and, related to that, to (3) discuss the role of *media organizations* and their responsibility for sustainable communication. This is related to their function to produce a "public good" with a certain *value*, which needs to meet the criteria of being "sustainable" and a contribution to social transformation processes as well.

## 3 Media, CSR, and Sustainability

The three dimensions will be briefly outlined and discussed in the following, followed by an overview of the contributions, collected in the volume at hand.

### *3.1 Sustainability as Guardrail for CSR (Media and Communicative Responsibility)*

As mentioned above, Corporate Social Responsibility (CSR) describes an established management focus of today's corporates and organizations of various kind, scope, and size. The social impact (SI) on the society and the key publics for which they function is lately debated in various fields of (mostly strategic) communication research (Rasche et al., 2017; Diehl et al., 2017; Allen, 2016; Heath, 2018; Hurst et al., 2020; Saffer, 2019). Alongside, the idea that organizations need the permission, the license to operate (SLO) (Hurst et al., 2020), challenges all kind of business, but media corporations in particular. Unlike CSR initiatives in other industry sectors, CSR and sustainability communication practices and related research in the media industry are still underdeveloped (Hou & Reber, 2011; Zhang & Swanson, 2006; Painter-Morland & Deslandes, 2017). Additionally, while even media corporations have started to think about their responsibility and begun to report on single activities, there seems to be a gap of criteria and guidelines for the "how" to do that. What is the normative framework for activities? Media freedom? Journalism ethics? Or sustainability as moral guidepost or compass or even the Sustainable Development Goals?

The challenges are that until recently the media industry has not been challenged to introduce sustainable and responsible business models anyway; furthermore, the watchdog-role that media play in observing traditional businesses and politics has provided a sufficient legitimacy for a long time—why to do more? Why to go beyond the role as the fourth estate? Lastly, the debate about the media's public value has covered questions about responsibilities toward the society and related impact so far.

In an era where fake news is constantly spread and algorithms co-decide the media agenda, the question about the impact on the public sphere, the public value of media products, and the license to operate are becoming prevalent with a new normative framework of sustainability. In this book we will bridge the "former" debate on *public value* with the current debate on *social impact* and the *social license to operate in the media industry*. In the focus is the double nature of producing economic and cultural goods at the same time (Bracker et al., 2017; Karmasin & Bichler, 2017) which leads to the assumption that media companies have a double responsibility for the way they present reality (in their products) and with this controlling and criticizing economic and political developments and raising ethical concerns in the public debate on the one hand (*social impact*), and for their own activities as a corporation on the other hand (*license to operate*).

Therefore, sustainability is not an object or key event that media can "report on." Instead, sustainability is a normative idea and stipulation to consider and anticipate the consequences of individual and organizational action. Sustainability is the guiding principle for individual and organizational action. Then, sustainability communication is all communication about specific issues (social, environmental, cultural, and economic issues), which thematizes, problematizes, discusses, and negotiates the principle of sustainability (Weder, 2021b).

Accordingly, media organizations are responsible to facilitate these negotiation processes, to enable sense- and meaning-making (Berger, 2009); they can create a culture of sustainability (Weder, 2022) by bringing together all aspects of sustainable action (Weder, 2022; Soini & Dessein, 2016). This process includes all facets of media actions and interventions that either promote communication for social change or media and product development and further institutionalization, including a formal policy on media sustainability.

## 3.2 Sustainability Communication as Part of Media CSR Concepts (Global Perspectives)

There are a few examples to look at where media corporations deal with their *twin responsibility* of holding society responsible and being responsible themselves, what media outlets perceive as their responsibilities and where they already action this with what kind of resources—and related to which normative principles. The Austrian Public Broadcaster ORF is one of the media organizations, like the BBC, where the Public Value debate is leading to a wider normative framework of a sustainable future, and related activities are sought to make the organizations accountable for their behavior—as a business, as a corporate citizen and as the producer of a critical public and "culture" (Weder, 2022).

As mentioned above, CSR & the media is a relationship where more research—and voices from the practice—is still needed. However, sustainability always adds another tier of complexity to any assessment—because it is not only "blurry" if used as master-frame (Weder, 2021a), but furthermore, it cannot be observed directly and must be estimated based on things that can be observed—but only retroactively. Additionally, sustainability relies on very individual perceptions and conceptions of what the future is and how it should look like—which makes it strongly related to the cultural context (O'Riordan & Voisey, 1997).

As a consequence, it is important to differentiate between disciplinary lenses, i.e., if media organizations and their responsibility are analyzed from an economic perspective or if we're looking at the sense- and meaning-making processes and the "value" in terms of the public and public interest from a media and communication studies perspective. Also, the cultural context is important, as is the media system (Hallin & Mancini, 2011; etc.; Möller et al., 2019). Cultural differences do not only exist in terms of media systems and societal value frameworks, but also in the way CSR is established, institutionalized, and/or even operationalized as a concept and sustainability and sustainable development are known and/or applied and referred to in the CSR activities.

## *3.3 Media Organizations, Public Value, and Sustainable Communication (Practical Insights)*

Not only global perspectives are needed, to understand all the dimensions of media corporate responsibility. The dimensions of responsibility have also to be reflected on considering innovative concepts of news production. The media industry needs a stronger focus on sustainability as normative framework and principle of action, as a moral compass for leadership, management, and journalism in media corporations, and needs to discuss how the CSR and sustainability debate is increasingly converging with the former public value debate.

Media corporations are not only measured by their financial value, as described above, they do not (only) get their legitimacy via market acceptance. They create an extra value, which is related to the dimension of the society as such, the so-called public value. This is not only related to what they do themselves and to their product, but media are also increasingly seen as important stakeholder, as an organization with which other organizations, mainly corporates, get engaged with as part of their CSR management—seeking "publicity," seeking to create awareness for their CSR activities (Tench et al., 2007). Here, public service media play a different role compared to private media corporations, who are much more depended on financial partnerships. Also, the European perspective with a historically strong role of public broadcasting can be contrasted with an Australian or US-perspective, with a much more concentrated and economically driven media landscape. This accentuates the importance to reframe the concept of public value with sustainability as potentially new "universal value."

In the following, a brief overview over the contributions to this volume is given, outlining their main topics, perspectives, and key insights related to questions around media CSR, responsible media management, and the role of sustainability as moral compass and universal "public value."

## 4 Contributions

The volume starts with three essays, which approach media corporate responsibility and sustainable communication from different perspectives.

Matthias Karmasin develops a framework that helps media outlets to communicate responsibility across cultural differences. He argues that in general a cultural transition calls for adaption of values and standards. But what impact does that notion have on an entity like CSR that mainly consist of values and standards? Karmasin offers criteria like respect to arguments by others, rational argumentation and—among others—openness and self-reflection as fundamental norms to path the way to communicate with integrity and about integrity.

A deeper understanding of the relationship between culture and sustainability is suggested by Franzisca Weder. Media corporates are discussed as responsible to cultivate sustainability in their organization and products. The idea discussed in the conceptual essay is that media have the responsibility to facilitate public discourses about and around sustainability, they need to create, facilitate, and maintain conversational spaces, where sustainability is negotiated as potential "universal value," and where examples are given for behavior and activities that are guided by sustainability as moral compass.

Rademacher concludes with linking these thoughts to theories of transformation and change. He proposes a "new deal" for news media that not only integrates classic elements like sensemaking and aspirational talk from organizational communication and phenomena like solution journalism but proposes a general ethos of sustainability across their media business models.

In the second part of the volume we take up a management perspective that asks how sustainability can work as a moral compass, and more specific as guardrail and framework for CSR and how the claim to act responsibly can be maintained even under critical conditions and in crisis situations.

In their chapter, Paul Clemens Murschetz, Eduard Frantz, and Niko Alm critically discuss the validity and performance measurement of public value theory. By looking at how Austria's main media player ORF is currently transforming its public service broadcasting, they present real-life insights to refine and reframe the concept of public value. After introducing the original public value concept and comparing it to the latest developments in the academic and public discussions, they argue that we need alternative quality-based indicators in order to measure Public Service Media

(PSM) and their activities effectively, e.g., through audience satisfaction and approval ratings or "value for money" measures. Finally, they discuss the general concept of finance models for public service media. They argue that even if the established model of license fees still dominates academic discourses, it needs to be fundamentally reviewed since the deficits of the model seem to outweigh its benefits. Besides new financing models, public value for PSM is a clear definition of its mission and vision; however, the impact for society needs to be defined and measurable.

The case study on the "Relotius scandal" involves Germany's most prominent news magazine DER SPIEGEL. Christian Pieter Hofmann and Stephan Russ-Mohl focus on communication efforts of this specific media company in a crisis situation, where—to their observation—the lines between journalism and corporate communications start to vanish. The authors describe CSR as a possible distinction when it comes to media coverage in a crisis narrative. Likewise, journalistic quality is perceived as an element of media companies' corporate responsibility. With this conceptual background they unfold the Relotius scandal where a highly respected journalist created media products almost too good to be true and managed to pass all editorial quality checks. The following internal investigation was perceived mostly as transparent, credible, and comprehensive by the interviewed experts. However, they found that through a joint crisis response team also editorial staff members repeated the corporate position and took the role of spokespersons—which under regular circumstances would be seen as a threat to editorial independence and journalistic quality.

Gerrit Boehnke analyzes the role of CSR in the recruiting communication of German and Austrian media brands. He merges the discussion on the importance of CSR for employee attractiveness with the debate on its value for overall business performance. His empirical analysis of web content of selected media companies highlights that CSR is not only important for B2C, but also important for B2B as well as corporate–employee relationships. However, he finds that CSR is not explicitly addressed within the career websites as a decision criterion or field of communication. This remarkable low-level of integration and relevance of CSR in the analyzed career websites contradict the idea that media companies explicitly aim at junior executives through CSR communication. Further research in this area might bring new insights into this contradiction.

Digital technologies challenge old values of media corporates. Cinzia Dal Zotto, Afshin Omidi, and Esmaeil Norouzi focus in their study on an effect that they call "smart exclusion": on the one hand, digital technologies could be harnessed to extract, produce, analyze, and distribute data in an intelligent way and thus allow news organizations to reach both profit and positive social impact. On the other hand, however, these technologies transform the nature of news work, moving its focus from creating public-informing ideas to distribution and therefore a focus on marketing activities. However, particularly smart technologies create a new, uncontested reality for defining the workplace in news organizations and demand mainly, if not only, the involvement of tech-savvy talents ideally best suited to work within this new context. As a result, "smart exclusion" emerges, which impacts on news organizations' social license to operate. This chapter conceptualizes the process of smart exclusion in news media, suggesting a symbiotic perspective for dealing with the related challenges and ensuring the respect of the diversity and inclusion dimensions of news organizations' social responsibility as part of their sustainable development.

In the third part of the volume, we take a deeper look into developments from a global perspective. We first examine the German development in recent years. Annkathrin Clemens and Josephine Franz compare CSR strategies on public and private media companies in Germany, interviewing media managers from the most important German outlets. The findings of a series of explorative interviews show that most of the managers identify their societal responsibility through their editorial content, i.e., public value. However, it became clear that overall the awareness for corporate responsibility and the need for generating social impact is growing and many media companies are actually developing specific CSR strategies. From a leadership perspective, media companies are increasingly confronted with growing legitimation pressure and high expectations of their stakeholders. Thus, they intend to increase their credibility with their CSR strategies. A strongly changed competitive environment and changed expectations from the applicant market require media companies to deal with CSR and take action, which has also been debated in Boehnke's contribution.

Moving over to Switzerland, Sarah Marschlich and Daniel Vogler analyze whether and if so, how news coverage about Media Social Responsibility (MSR) has changed over time, how news coverage about MSR contributes to the media reputation of media organizations, and how these aspects differ between public and private media organizations and owned and non-owned media outlets. They conducted a quantitative content analysis of news articles in 10 Swiss media outlets covering the social aspects of Swiss media organizations from 2010 to 2019. Overall, their results show that MSR coverage has increased over the period, primarily driven by coverage of public media organizations' MSR. However, the tone of the articles was mainly negative, implying that MSR may pose a risk to the social reputation of media organizations. By providing comprehensive insights into the role of the media in negotiating MSR, their study significantly contributes to the emerging field of MSR research.

Tevhide Serra Gorpe and Burku Öksüz aim to explore the concept of media responsibility and CSR of media institutions in Turkey from a journalism educators' points of view. Their contribution focuses on gaining a better understanding of journalism educators' perceptions of *responsibility*—with respect to the Turkish media landscape. The research has practical implications on journalism education on a broader level because it investigates how and to what extent issues like CSR, sustainability, or social impact are integrated to curricula, including the challenges in teaching media corporate responsibility to journalism students. The authors show that Turkish educators in particular see the main task of media corporate CSR facilitated through journalism, and thus in writing good and objective news—and not in setting up CSR strategies and projects. However, from an education perspective, the authors critically reflect that media partnering on CSR matters with NGOs and other entities which is recommended by the interviewed journalism educators might not be enough in a changing news world and for creating social impact.

Bringing in a perspective from South America, in her chapter Elena Block discusses the viability of news media's corporate social responsibility (CSR) programs in Venezuela—and thus, in a different political and media landscape compared to the previously discussed European examples. She describes the politicization of media, their strategies, and this an ongoing "government–media clash" which has also dominated the country's political communication for two decades. By exploring how the deep political polarization, the mediatization of politics, and politicization of the news media effect the CSR programs of media companies she shows that despite the successful media and journalistic education initiatives the most important CSR work is not the corporate "program," but rather the everyday journalistic work which is still following the idea(l) of providing trustworthy, ethical, community-engaged, public service journalism. The chapter shows, that in a country like Venezuela, the social impact and Corporate Social Responsibility is rather "just" covering the daily news and stories that are of real importance to the people and communities, thus, public media serves the interests of society and makes by doing this a real impact to people's life.

Richard Murry writes about "Post-Murdoch" regional media in Australia where news media are increasingly under pressure. Looking especially at the disrupted media scene in Queensland where the Star Group filled the void that was left by Murchdoch's News Corp. pulling out of the regional newspapers they owned after the COVID-19 pandemic, he shows how journalist work beyond their contracted working hours. That leads to a situation where journalism is shallow and predictable that the conclusion is that only "the illusion of journalism exists": The news media are still present but they are unable or unwilling to take on stories that challenge authorities or support democracy. Murray's gripping chapter reminds us about the fundamental value of free and independent journalistic systems and draws a rather dystopic scenario of what is to come in other western countries.

From this chapter, we transition to a number of practical insights on sustainability related activities in specific media corporations. We start with Peter Greste's rather skeptical vision of what he calls the infodemic. In his essay he proposes the death of objectivity but still leaves us with the optimistic notion that journalism has a chance to save his objectivity. Second contribution is delivered by Wolfgang Reising, who explains that Public Service Media are perceived not only along their mission of securing independent news and information; he defines media companies as foremost economic actor who is confronted with a number of expectations; thus, CSR activities are a matter of compliance with these expectations. This view is also supported by Katja Bäuerlein from UFA, a private TV, and film producer. She explains the system of norms and believes which led to a complex CSR program. However, from her perspective the main impact is less on the society but much more on the employees—which supports some assumptions developed in the previous chapters of this volume. Finally, two interviews lead us into the heart of both Austrian and German Public Service Media, ORF and ZDF. Both public broadcasters started early with CSR reporting and activities, that are indeed directed toward the own employees, and here specifically the journalists. Anita Malli points to the trainings offered for journalists to be informed around climate change and sustainability related issues to use this lens in all aspects of their reporting. However, she emphasizes that the trainings need to be requested by the leadership, mainly the chief editors, and, beyond that, the general management.

Greste's contribution supports one of the main lines of argumentation of this volume. Looking into different countries, media systems and levels of responsibility, from individual, institutional, to professional and systemic responsibility, sustainability has the potential to be a guiding framework and moral compass for activities under the strategic umbrella of CSR. However, there is still the need to further negotiate the meaning of sustainability from a media and journalism perspective; related values like independence, objectivity need to be discussed in the same way as fairness, diversity or inclusion, and equality, which can be one step toward a new Sustainable Development Goal #18: sustainable communication and media.

## References


Allen, M. (2016). *Strategic communication for sustainable organizations. Theory and practice.* University of Arkansas.

Berger, G. (2009). How the internet impacts on international news: Exploring paradoxes of the most global medium in a time of hyperlocalism. *International Communication Gazette, 71*(5), 355–371.

Bracker, I., Schuhknecht, S., & Altmeppen, K. D. (2017). Managing values: Analyzing corporate social responsibility in media companies from a structuration theory perspective. In *Value-oriented media management.* (pp. 159–172). Springer, .

Christensen, L. T., Morsing, M., & Thyssen, O. (2017). License to critique: A communication perspective on sustainability standards. *Business Ethics Quarterly, 27*(2), 239–262.

Diehl, S., Karmasin, M., Mueller, B., Terlutter, R., & Weder, F. (Eds.). (2017). *Handbook of integrated CSR communication.* Springer.

Elving, W. J., Golob, U., Podnar, K., Ellerup-Nielsen, A., & Thomson, C. (2015). The bad, the ugly and the good: New challenges for CSR communication. *Corporate Communications: An International Journal, 20*(2), 118–127.

EU Greenbook. (2001). Source: https://www.jussemper.org/Resources/Corporate%20Activity/Resources/IMPLEMENTINGPARTNERSHIPJOBS.pdf.

Freeman, R. E., & Moutchnik, A. (2013). Stakeholder management and CSR: Questions and answers. *Uwf UmweltWirtschaftsForum, 21*(1), 5–9.

Golob, U., & Podnar, K. (2011). Corporate social responsibility communication and dialogue. *The handbook of communication and corporate social responsibility.* (pp. 231–251).

Golob, U., Podnar, K., Elving, W. J., Nielsen, A. E., Thomsen, C., & Schultz, F. (2013). CSR communication: Quo vadis? *Corporate Communications: An International Journal.*

Grunwald, A. (2012). *Sustainability assessment of technologies—an integrative approach.* IntechOpen.

Hallin, D. C., & Mancini, P. (Eds.). (2011). *Comparing media systems beyond the Western world.* Cambridge University Press.

Heath, R. L. (2018). Issues management. *The International Encyclopedia of Strategic Communication,* 1–15.

Hou, J., & Reber, B. H. (2011). Dimensions of disclosures: Corporate social responsibility (CSR) reporting by media companies. *Public Relations Review, 37*(2), 166–168.

Hurst, B., Johnston, K. A., & Lane, A. B. (2020). Engaging for a social licence to operate (SLO). *Public Relations Review, 46*(4), 101931.
Hurst, B., & Johnston, K. A. (2021). The social imperative in public relations: Utilities of social impact, social license and engagement. *Public Relations Review, 47*(2), 102039.
Johnston, K. A., & Lane, A. B. (2021). Communication with intent: A typology of communicative interaction in engagement. *Public Relations Review, 47*(1), 101925.
Karmasin, M., & Bichler, K. (2017). Corporate social responsibility in the media industry: Setting the benchmark or falling behind?. In *Value-oriented media management.* (pp. 135–146). Springer.
Möller, J. E., Nölleke-Przybylski, P., Voci, D., von Rimscha, M. B., Altmeppen, K. D., & Karmasin, M. (2019). A motivation-based typology of media companies' cross-border engagement. *European Journal of Communication, 34*(3), 300–318.
O'Riordan, T., & Voisey, H. (1997). The political economy of sustainable development. *Environmental Politics, 6*(1), 1–23.
Painter-Morland, M., & Deslandes, G. (2017). Reconceptualizing CSR in the media industry as relational accountability. *Journal of Business Ethics, 143*(4), 665–679.
Rasche, A., Morsing, M., & Moon, J. (Eds.). (2017). *Corporate social responsibility: Strategy, communication, governance.* Cambridge University Press.
Saffer, A. J. (2019). Fostering social capital in an international multi-stakeholder issue network. *Public Relations Review, 45*(2), 282–296.
Schoeneborn, D., & Trittin, H. (2013). Transcending transmission: Towards a constitutive perspective on CSR communication. *Corporate Communications: An International Journal.*
Soini, K., & Dessein, J. (2016). Culture-sustainability relation: Towards a conceptual framework. *Sustainability, 8*(2), 167.
Tench, R., Bowd, R., & Jones, B. (2007). Perceptions and perspectives: Corporate social responsibility and the media. *Journal of Communication Management.*
Trittin, H., & Schoeneborn, D. (2017). Diversity as polyphony: Reconceptualizing diversity management from a communication-centered perspective. *Journal of Business Ethics, 144*(2), 305–322.
UN. (2022). Sustainable development goals. Source: https://sdgs.un.org/goals.
WCED. (1987). Commissione mondiale per l'ambiente e lo sviluppo, Brundtland-kommissionen, Comisión Mundial del Medio Ambiente y del Desarrollo, United Nations. General Assembly, WCED., Weltkommission für Umwelt und Entwicklung, ... & Verdenskommisjonen for miljø og utvikling. (1987). Our common future. Peterson's.
Weder, F. (2021a). Sustainability as master frame of the future? Potency and limits of sustainability as normative framework in corporate, political and NGO communication. In *The sustainability communication reader.* (pp. 103–119). Springer VS, .
Weder, F. (2021b). Strategic problematization of sustainability reframing dissent in strategic communication for transformation. *Public Relations Inquiry*, 2046147X211026857.
Weder, F. (2022). Nachhaltigkeit kultivieren. Öffentliche Kommunikation über Umwelt, Klima, nachhaltige Entwicklung und Transformation. *Communicatio Socialis (ComSoc), 55*(2), 146–159.
Weder, F., Einwiller, S., & Eberwein, T. (2019). Heading for new shores: Impact orientation of CSR communication and the need for communicative responsibility. *Corporate Communications: An International Journal.*
Winkler, P., Etter, M., & Castelló, I. (2020). Vicious and virtuous circles of aspirational talk: From self-persuasive to agonistic CSR rhetoric. *Business & Society, 59*(1), 98–128.
Zhang, J., & Swanson, D. (2006). Analysis of news media's representation of corporate social responsibility (CSR). *Public Relations Quarterly, 51*(2), 13.

# Conceptualizing Media CSR Communication: Responsible Contributions to the (Global) Public Sphere?

**Matthias Karmasin**

## 1 Introduction: CSR Communication in the Media Industry

Communication is a global business. But it is not only a business. Every corporation is responsible for its actions and inactions, but media corporations do have a special responsibility, as they do not offer a simple service and just allocate economic goods but they provide a cultural good as well and contribute to the public sphere and sometimes constitute it. This special responsibility is acknowledged and communicated by many media corporations: the CSR debate has reached the media industry[1] and CSR as a management tool has arrived in many media companies—including the appropriate communication of CSR activities.[2]

As in other industries there is not universal and consensual definition of the term corporate (social) responsibility in the media industry. The approaches and definitions of CSR and the CSR activities of media corporations cover a wide range. Their CSR self-declarations are quite divers as they engage in CSR for many reasons: to maintain their license to operate, to increase brand value and reputation, to enhance trust (especially when journalistic content is addressed), to establish sustainable stakeholder relations, and to try to avoid regulation and of course as well to be

[1] E.g., Tench et al. (2007), Toker (2013), Karmasin et al. (2014), Batko and Kreft (2017), Bracker (2017), Ingenhoff and Bachmann (2017), Koinig et al. (2019), Trommershausen and Karmasin (2020).

[2] E.g., Gulyás (2009), Ingenhoff and Koelling (2012), Karmasin and Bichler (2017), Kalinowska-Żeleźnik et al. (2017).

M. Karmasin (✉)
CMC Austrian Academy of Science/University of Klagenfurt, Vienna, Austria
e-mail: matthias.karmasin@aau.at



F. Weder et al. (eds.), *CSR Communication in the Media*, CSR, Sustainability, Ethics & Governance, https://doi.org/10.1007/978-3-031-18976-0_2

good corporate citizens and to act responsibly.[3] Even if the societal relevance of content (e.g., responsible entertainment) and the role of the media as fourth estate and the responsibilities for the quality of the public sphere are addressed in many CSR statements the industry in general is not setting CSR benchmarks (Karmasin & Bichler, 2017): neither when it comes to ethical self-reflection and responsible self-limitation of business opportunities nor when it comes to communication of CSR (e.g., Gulyás, 2009, 2011; Bracker, 2017). In other words the critics of unethical behavior in other sectors of society are apparently not willing to be measured by the same (or even stricter) standards themselves—this seems to hold true for legacy media and for "social" media and intermediaries alike. So, the "polyphony of CSR" (Castello et al., 2013) and thus the "polyphony of CSR communication" (Karmasin & Apfelthaler, 2017) also fully apply to the media industry.

In this chapter this polyphony is not reconstructed. The differences in definitions and self-declarations of CSR within the media industry and as compared to other sectors, the changes in CSR concepts due to sustainability as an emerging topic and the levels of adaption to the SDGs, and the differences between academic and political definitions of CSR and the actual concepts in the corporate world are addressed in many academic reflections—also in this volume.

This contribution focuses on the polyphony of CSR communication and especially on the aspect of communicating CSR across cultures and on the ethical implications and challenges in and of these processes—making the difference between is and ought count in this context as well. It may seem that this is a rather minor detail, but the translocal (global) nature of the climate and pandemic crisis, the public spheres in which these topics and issues are discussed, and the role and responsibility international media corporations play in the allocation of content therein may serve as examples for a rising importance of addressing these issues.

## 2 Relativism vs. Universalism: Communicating CSR Across Cultures

Under the premises of globalization, digitalization, and mediatization the old "global versus local dilemma" (Jamali, 2010) is more valid than ever for media corporations. Their core business is content sourcing, (2) content aggregation, and (3) content dissemination[4] and the allocation of content is technically possible via an almost global (and digital) infrastructure. But not socially and culturally, as content is embedded in culture as Voci (2019) points out and local adaptions are essential especially when it comes to news. But a language wise adaption is often not enough—simple translation does not do the job. Many differences play a role as

[3]E.g., Ingenhoff and Koelling (2012), Kalinowska-Żeleźnik et al. (2017).

[4]This definition would include "intermediaries" and define them as media companies—but in times of convergence this of course is an ongoing debate—see: Voci et al. (2019).

the degree of mediatization and divergent media usage practices, education and media literacy, different religion and ethical norms, the overall economic situation, and in particular the structures of media refinancing (advertising and the willingness to pay for content) the political structure and media regulation and governance to name just a few.

Culture matters not only for the allocation and refinancing of content itself but especially when it comes to licenses to operate and the definition of good corporate citizenship: a good corporate media citizen, e.g., in China, Poland, Germany, Hungary, England or in Australia is defined quite differently.[5]

So the polyphony of CSR communication is not only a consequence of divergent corporate cultures, but also seems to root in different cultural settings and different cultural norms. According to Gulyas (2011), these differences can exist in (1) the format of CSR communication, (2) the extent of CSR communication, and (3) the content of CSR communication (Table 1). In order to structure this polyphony and condense the empirical variance, Karmasin and Apfelthaler (2017) suggested three prototypes to CSR communication across cultures based on various empirical studies[6]:

From an instrumentalistic and strategic viewpoint with a focus on efficiency a geocentric (with a severe loss in standardization) or a polycentric approach seems promising, as the conflict between values is not addressed neither in the format of CSR communication nor in the extent of CSR communication, and not in the content of CSR communication as well—and the content presented (and/or allocated) in the different cultural settings not only follows the formal requirements (as different media regulation settings) but also shared cultural beliefs and normative settings in the host country. In case of a polycentric approach the corporation does not impose any values upon other cultures and reacts flexibly to different moral beliefs and ethical practices in different settings. If the different values do not conflict, if there

---

[5] For Europe, e.g., https://ec.europa.eu/info/policies/justice-and-fundamental-rights/upholding-rule-law/rule-law/rule-law-mechanism/2021-rule-law-report/2021-rule-law-report-communication-and-country-chapters_en.

[6] For China (Liu et al., 2014; Luethge & Han, 2012), a study involving data from 49 countries (Ho et al., 2012), Denmark (Morsing et al., 2008), Germany (Antal et al., 2009), Hungary (Ligeti & Oravecz, 2008), India (Chaudhri & Wang, 2007), Japan (Takano, 2013), Mexico (Meyskens & Paul, 2010; Logsdon et al., 2006) Korea (Kim and Kim, 2010) or Russia (Bashtovaya, 2014; Soboleva, 2007) and comparative studies such as, for instance, those comparing Russia and the United States (Bashtovaya, 2014), Brazil and the United Kingdom (Abreu & Barlow, 2013), Dutch and India (Planken et al., 2013) Germany, Italy, and the United States (Habisch et al., 2011), the United Kingdom, France, the United States, and Canada (Freeman & Hasnaoui, 2011), Canada, the United Kingdom and Germany (Bondy et al., 2004, 2008) or France, the Netherlands, the United Kingdom, and the United States (Maignan & Ralston, 2002), US, UAE and South Korea (Rim & Dong, 2018) The explicit approach to CSR that is followed in the United States also often results in more deliberate CSR communication, whereas the implicit European approach goes hand in hand with poor CSR communication (Schmeltz, 2014; Habisch et al., 2011). For South Korea and USA in regard to Carroll's four-dimensional model of CSR Kim et al. (2019).

**Table 1** Strategic approaches and CSR communication outcomes in different environments

| | Dominant strategy | Approach to CSR communication in host country is *similar* to dominant perspective or standard | Approach to CSR communication in host country is *different* from dominant perspective or standard |
|---|---|---|---|
| *Ethnocentric* | *Unilateral standardization*: home country perspective or standard alone. Universalistic ethical approach | *Match.* Home country approach to CSR communication applied in host country<br>Result: no conflict of culturally biased perspectives; global efficiency in CSR communication through standardization | *Mismatch.* Home country approach to CSR communication applied in host country<br>Result: conflict of culturally biased perspectives; entering MNE perceived as foreign imperialist with CSR communication standards and practices that do not fit the local environment |
| *Polycentric* | *Bilateral adaptation*: home country perspective or standard exists, but host country perspective or standard embraced. Dialogic ethical approach | *Match.* Host country approach to CSR communication applied in host country<br>Result: no conflict of culturally biased perspectives; over-adaptation to meaningless differences may lead to fragmentation of CSR communication | *Non-consequential mismatch.* Host country approach to CSR communication applied in host country<br>Result: no conflict of culturally biased perspectives; fragmentation of CSR communication due to necessary adaptation |
| *Geocentric* | *Global standardization*: no home country perspective or standard. Global perspective or standard established Relativistic ethical approach | *Match.* Global approach to CSR communication applied in host country<br>Result: no conflict of culturally biased perspectives; global efficiency in CSR communication through standardization | *Mismatch.* Global approach to CSR communication applied in host country<br>Result: conflict of global vs. local perspectives; entering MNE perceived as global imperialist imposing CSR communication standards and practices that do not fit the local environment |

are no legal restraints and the local values do not harm the corporate or brand identity, this may seem efficient and plausible.

But as Szőcs and Miska (2020) summarize managing across cultures implies ethical challenges beyond the mere efficiency. Especially for the communication of CSR the question is not "effectiveness" or "success" but also if the communication in itself is responsible (as, e.g., Weder & Karmasin, 2017 have pointed out). To discuss an example: what if a media corporation coming from a home country with democratic values, equal opportunity ideals and freedom of speech guaranteed in the constitution, with not interventionist media regulation and a high degree of professional media accountability operates in host countries where there is no freedom of speech (even censorship), no media freedom, and a social environment in which women are not only not treated equally but where this is a widely accepted norm. There are media environments, in which such practices are widely accepted, and—being part of the cultural heritage—they are supported by local laws and by practical customs. In such an environment, should the media corporation still "do as the Romans do (when in Rome)"? Should it just address the relativity of CSR in these settings or not communicate CSR at all there? Or should the media enterprise follow an ethnocentric approach and be a global leader and accept the relevance of content and the interconnection of CSR and ethics and meet the challenges of responsible global leadership as outlined by Stahl et al. (2016):

1. Diversity: The need to address the legitimate claims, rights, and expectations of a diverse group of stakeholders, both within and outside the organization.
2. Ethics: The need to ensure principle-driven, legally sound, and ethically acceptable behavior.
3. Sustainability: The need to contribute to protecting, preserving, and restoring the resources of the environment while improving human wellbeing and social equity.
4. Citizenship: The need to recognize, understand, and effectively address economic and social issues with respect to equality, social justice, and human rights protections.

One might argue that especially media corporations—given that their key business model rests on (1) content sourcing, (2) content aggregation, and (3) content dissemination—and that this contributes to and sometimes constitutes the public sphere—do have a special responsibility to lead globally in terms of ethical responsibility. Thus a strict—"do as Romans do when in Rome" approach is not appropriate from an ethical viewpoint neither in their CSR statements nor in their CSR communication and that they should address universal standards (like human rights and human dignity) and promote "legitimate claims, rights, and expectations of a diverse group of stakeholders" and to "recognize, understand, and effectively address economic and social issues with respect to equality, social justice, and human rights protections." Of course this arguments rests on the assumption that there are universal standards—for ethics and for CSR—valid across cultures and countries.

A number of global initiatives as the universal declaration of human rights, the Global Reporting Initiative (GRI), the Fortune 100 Global Accountability List, the Account Ability's AA 1000 Responsibility Assurance Standard, the FTSE4GOOD, the Dow Jones Sustainability Index, the UN Global Compact's COP (based on the "Ruggie Principles"), Social Accountability International's SA8000 standards, CSR Hub's performance ratings, or even the ISO 26000 CSR or lately working toward global sustainability the SDG's do indicate this.

Following this line of argumentation—and the idea of CSR as self-limitation (even if it reduces business opportunities) and the idea of consistent CSR communication—it seems appropriate for a media corporation to clearly define the standpoint of the corporation (on human rights, freedom of opinion and speech, ethical standards in relation to the development of the host country, to the SDG's, to local taboos, and so on) and to communicate it thereafter. Maybe even to make this declaration part of the editorial guidelines, the automatized selection of content (via AI), of the production of entertaining content and even to incorporate it in (journalism) trainings. Standardization therefore seems to be the ethically justified answer—one set of values entrenched by universal declarations and statements, consistently communicated, integrated across all departments and subsidiaries, embedded in a consistent branding and corporate strategy, enforced with rigorous quality control, trained, and integrated in the accountability practices.

But: what if the CSR communication conflicts with local taboos and communication practices, and what if the communication of these set of values itself is regarded as inappropriate and unacceptable by the host country? Or even illegal—as the call for equal opportunity and for freedom of speech is not welcomed alike globally. What if the media corporation would be banned from the host country or managers, editors, programmers, and journalists would be threatened with lawsuits, violence, and even imprisonment? Should they still stay on the CSR message? And what if the argument is valid, that the message in itself is problematic from an normative viewpoint as there is globalization but no ethical universalism?

Even if globalization is a fact all of us still do live in a pluralistic world and thus do not share the same moral and ethical standards. The MIT moral machine experiment with decisions from 70,000 participants from 42 countries clearly shows a great variance in moral decisions. (Awad et al., 2018, 2020). Apart from this (and other) empirical evidence even the normative and (utopian) notion of universalism has been called in to question most prominently lately by Tingyang (2020) based on Confucian optimality[7] and relational rationality and a criticism of Western shortcomings in thinking traditions like Christianity's dualism, Hegel's dialectic and of the (utilitarian) idea of maximizing self-interest. What if as Adi and Stoeckle (2021) put it CSR and sustainability are a 'First World Problem'? What if the "weirdest people in the world" (Henrich, 2020) are substantially biased and wrongly universalize their particularity[8] and thus do not have the right to impose their preferences and ethical beliefs on others—also via media corporations. This shortcoming in regard to cultural sensitivity is not compensated by participating in

[7]Confucian improvement is more reciprocal and focused on collective prosperity than "Pareto improvement," which requires that no one gets worse when someone else gets better. Here, too, the global pandemic is a case in point: with a contagion like the coronavirus, no one is safe unless everyone is safe.

[8]As Henrich (2020) argues also via massively biased samples in empirical studies (with a bias towards the global north). "Western samples typically anchored the extreme end of the distribution. They were psychologically weird." (Henrich, 2020:xiii)

media accountability instruments in the host countries.[9] This is highlighted by the example of Western media companies which expanded to Eastern Europe in the last decades. These expansions were accompanied by the implementation of their corporate culture, especially the Western concepts of media accountability (Fengler et al., 2011, p. 12)—without much (cross-) cultural sensitivity.

And what if CSR communication in an ethnocentric approach is seen as a Trojan horse for cultural imperialism and global mainstreaming of content and thus is a form of unjustified normative imperialism.

These questions are somehow disturbing (at least from the perspective of the global north) as they call the assumption that there are universal ethical standards into question. What if there are no universal ethical principles, is there any right or ethical legitimation to interfere with other lifestyles and ethical beliefs? Is an polycentric approach thus the only normative concept to follow? Especially when the core business is content?

Such a point of view, however, would justify everything if it is just common practice in a given culture, even regarding the media just as an institution for maximizing capital income or promoting the image of those in power, or accepting censorship as an euphemism (e.g., as avoiding unrest through dissenting opinion and fostering peace in society). This of course does not only conflict with the set of values in the home country in the brief example, but could also endanger the license to operate in the home country and damage the reputation of good (global) citizenship. For example, by a loss of trust by being accused of double standards—in the home country and the host countries—a delicate topic when it comes to ethical responsibilities and the credibility of CSR communication—especially in relation to journalistic professionalism, media accountability, and content allocation.

To make a long story short (hopefully not too short): The old dilemma of universalism vs. relativism in cross-cultural ethics, the question of ethical leadership, the polyphony of CSR communication also applies to media corporations. The line between a local taboo with little ethical relevance (e.g., in case of certain food taboos) and the respect for other cultures in these matters and an irresponsible relativism (violating human dignity) is thin and blurry—but at least the standpoint the corporation takes should be clarified. Not to communicate CSR at all does not do the job: especially for companies whose core business is content. There is no easy way out—media corporations do have to take responsibility also in these matters and they have to continue the balancing act between a geo—or ethnocentric, standardized and a polycentric, decentralized approach—in all aspects of their CSR communication: form and content alike.

[9] For an overview, Eberwein et al. (2018, 2019).

## 3 Contributions to the Global Public Sphere? CSR Communication as Stakeholder Communication

The fundamental approach to CSR communication, not only to communicate about responsibility, but also to act responsibility through communication, as Weder and Karmasin argue (2017), has special relevance for the CSR communication of international media corporations as outlined above. Mainly because these corporations do not provide a simple service and do not allocate a trivial good—but as they contribute to the public sphere and tend to inform, and entertain societies—and thus they are responsible for value statements, public images and values, and the quality of the public sphere (e.g., by avoiding false balance, by pointing out partisan interests, by hindering manipulation and misinformation, by acting as the fourth estate). In terms of the content of CSR communication of multinational media corporations, these declarations should also reflect the question of universalism and relativism. Their CSR communication should not simply react to cultural differences but should also try to communicate about the respective cultural standards and how they could be ethically justified. Following Karmasin and Apfelthaler (2017) this does not imply that CSR communication has to impose the ethical standards of the home country to the host country (the problems of a strictly universalistic perspective have been argued above), but that the self-imposed CSR standards should be not only properly communicated but also ethically justified. CSR communication should always be communication with integrity. Or, to put it differently, business integrity is a basis for communicating with integrity.[10] The CSR communication strategy and the overall CSR strategy should provide the framework for communicating with integrity.

This means, first of all, that the CSR (communication) strategy in international dimension makes clear that it is only admissible to "do as the Romans do when the Romans do the right thing", and that it has to define what the right thing and circumstances are, and to justify this from an ethical (and not only strategic) point of view. As outlined above the lines between avoiding cultural imperialism and respecting local taboos and an irresponsible relativism are sometimes blurry and need argumentation and refinement—and that different content production and dissemination models and divergent ways of refinancing also imply different ethical challenges. The problems of intercultural communication should be addressed in the CSR communication processes itself and the premises of CSR and the underlying ethical arguments should be reflected—in terms of media as well as business ethics and with a special focus in the responsibility and accountability of media corporations and their relation to the quality of the public sphere—in the home and the host country and globally.

As outlined above for the communication of CSR the most important question is if the communication in itself is responsible, also in the form of the communication.

[10] As Schlegelmilch and Pollach (2005) point out, any successful communication of corporate ethics has to be grounded on a firm and sustainable commitment to responsibility and ethical behavior.

Following Stahl et al. (2016) the need to address the legitimate claims, rights, and expectations of a diverse group of stakeholders, both within and outside the organization should be reflected in the international CSR communication of media corporations.

On a formal level, the need for CSR communication (in the meaning of taking responsibility seriously) and the goal of integrating stakeholders in communication processes should be part of the CSR communication strategy. The integration of stakeholders in communicative processes, however, depends on the technical possibilities and social realities of a culture. "With digitalization, the parameters of relationships between organizations and stakeholders have changed fundamentally" (Lock, 2019)—also for media corporations.

Moreover: especially media corporations do have special and unique opportunities for stakeholder dialogue—and for promoting stakeholder dialogue and stakeholder communication also via social media and new forms of collaborative content platforms.[11]

Following Karmasin (2002), Karmasin and Apfelthaler (2017) and the idea of process ethics (see Krainer, 2002 or Heintel et al., 2006) the following set of norms is fundamental for this approach[12]:

- respect for arguments brought forward by others;
- rational argumentation (at least to refrain from the temptation to destroy those that cannot be convinced);
- openness and self-reflection (in order to establish transparency and openness);
- the intention to reach a universal position;
- and as this is often not possible: the will to agree to differ peacefully.

CSR communication in this understanding could also be an instrument to "reconceptualize CSR in the media industry as relational accountability" as Painter-Morland and Deslandes (2017) argue from an empirical perspective.

The idea of stakeholder communication and a dialogical ethical approach to CSR communication could generate an adaptive way of CSR communication, trying to balance the need for universalism and relativism, the necessity of norms and commitment, the credibility, and the trustworthiness of CSR communication.

Not only to communicate *about* integrity, but also to communicate *with* integrity should be the guiding principle for the CSR communication of media corporations and if they do not set the benchmark in ethical standards in their CSR communication content and their content in general—they could at least set standards in the form of their CSR communication and thus contribute to a global public sphere by

[11] E.g., Saxton and Waters (2014), Moreno et al. (2015), Manetti et al. (2017), Navarro et al. (2017), Watkins (2017), Men et al. (2018), Lai and Fu (2020), Wang and Yang (2020) on stakeholder communication via social media.

[12] This chapter does not have the room to engage in the debate on hypernorms (norms creating norms) in ethical terms, but it seems important to stress the argument that only hypernorms can avoid cultural biases in ethical arguments—so the key seems to be the process and the dialogue and not the convincing materiality of the argument in itself.

promoting a dialogical and stakeholder oriented way of CSR communication—and maybe at least to promote the idea of agreeing to differ peacefully globally. It may seem that this is a rather limited utopian horizon for international and global media corporations and their CSR communication—but given the status quo also in regard to the climate crisis and the (misinformation) pandemic it also could be seen as a massive improvement.

## References

Abreu, M. C. S., & Barlow, C. (2013). A comparative picture of corporate social responsibility approaches by leading companies in the United Kingdom and Brazil. *Social Responsibility Journal, 9*(4), 571–588.

Adi, A., & Stoeckle, T. (2021). Are CSR and sustainability a 'First World Problem'? Western and eastern European perspectives. In F. Weder, L. Krainer, & M. Karmasin (Eds.), *The sustainability communication reader: A reflective compendium* (pp. 567–584).

Antal, A. B., Oppen, M., & Sobczak, A. (2009). (Re-)discovering the social responsibility of business in Germany. *Journal of Business Ethics, 89*, 285–301.

Awad, E., Dsouza, S., Kim, R., et al. (2018). The moral machine experiment. *Nature, 563*, 59–64.

Awad, E., Dsouza, S., Shariff, A., Rahwan, I., & Bonnefon, J.-F. (2020). Universals and variations in moral decisions made in 42 countries by 70,000 participants. *Proc Natl Acad Sci U S A, 117*(5), 2332–2337.

Bashtovaya, V. (2014). CSR reporting in the United States and Russia. *Social Responsibility Journal, 10*(1), 68–84.

Batko, R., & Kreft, J. (2017). The superfluousness of CSR in media organisations: Conclusions form public radio research in Poland. *Journal of Organizational Change Management, 30*(1), 91–105.

Bondy, K., Matten, D., & Moon, J. (2004). The adoption of voluntary codes of conduct in MNCs: A three-country comparative study. *Business and Society Review, 109*(4), 449–477.

Bondy, K., Matten, D., & Moon, J. (2008). Multinational corporation codes of conduct: Governance tools for corporate social responsibility. *Corporate Governance: An International Review, 16*(4), 294–311.

Bracker, I. (2017). Corporate social responsibility (CSR) und corporate citizenship (CC): Selbstbild und Fremdwahrnehmung in der öffentlichen Kommunikation. In K. D. Altmeppen, F. Zschaler, H. M. Zademach, C. Böttigheimer, & M. Müller (Eds.), *Nachhaltigkeit in Umwelt, Wirtschaft und Gesellschaft* (pp. 257–284). Springer VS.

Castello, I., Morsing, M., & Schultz, F. (2013). Communicative dynamics and the polyphony of corporate social responsibility in the network society. *Journal of Business Ethics, 118*, 683–694.

Chaudhri, V., & Wang, J. (2007). Communicating corporate social responsibility on the internet a case study of the top 100 information technology companies in India. *Management Communication Quarterly, 21*(2), 232–247.

Eberwein, T., Fengler, S., & Karmasin, M. (Eds.). (2018). *The European handbook of media accountability*. Routledge.

Eberwein, T., Fengler, S., & Karmasin, M. (Eds.). (2019). *Media accountability in the era of post-truth politics: European challenges and perspectives. ECREA-Routledge studies in European Communication Research and Education*. Routlegde.

Fengler, S., et al. (2011). Mapping media accountability-in Europe and beyond. In T. Eberwein, S. Fengler, E. Lauk, & T. Leppik-Bork (Eds.), *Mapping media accountability-in Europe and beyond* (pp. 7–22). Herbert von Halem Verlag.

Freeman, I., & Hasnaoui, A. (2011). The meaning of corporate social responsibility: The vision of four nations. *Journal of Business Ethics, 100*, 419–443.

Gulyás, Á. (2009). Corporate social responsibility in the British media industries—preliminary findings. *Media, Culture & Society, 31*(4), 657–668.

Gulyas, A. (2011). Demons into angels? Corporate social responsibility and media organisations. *Critical Survey, 23*(2), 56–74.

Gulyás, Á. (2011). Demons into angels? Corporate social responsibility and media organisations. *Critical Survey, 23*(2), 56–74. https://www.jstor.org/stable/41556417. Accessed 6 April 2021

Habisch, A., Patelli, L., Pedrini, M., & Schwartz, C. (2011). Different talks with different folks: A comparative survey of stakeholder dialog in Germany, Italy, and the U.S. *Journal of Business Ethics, 100*, 381–404.

Heintel, P., Krainer, L., & Ukowitz, M. (Eds.). (2006). *Beratung und Ethik. Praxis, Modelle, Dimensionen*. Ulrich Leutner Verlag.

Henrich, J. (2020). *The WEIRDest people in the world: How the west became psychologically peculiar and particularly prosperous*. Macmillan.

Ho, F. N., Wang, H. M. D., & Vitell, S. J. (2012). A global analysis of corporate social performance: The effects of cultural and geographic environments. *Journal of Business Ethics, 107*, 423–433.

Ingenhoff, D., & Bachmann, P. (2017). How do media companies gain legitimacy? An experimental study on the (ir)relevance of CSR communication. *International Journal of Strategic Communication, 11*(1), 79–94.

Ingenhoff, D., & Koelling, A. M. (2012). Media governance and corporate social responsibility of media organizations: An international comparison. *Business Ethics: A European Review, 21*(2), 154–167.

Jamali, D. (2010). The CSR of MNC subsidiaries in developing countries: Global, local, substantive or diluted? *Journal of Business Ethics, 93*(2), 181–200.

Kalinowska-Żeleźnik, A., Kuczamer-Kłopotowska, S., & Lusińska, A. (2017). The reflections of the CSR strategy in the activities of a public medium, as exemplified by radio Gdańsk. *Journal of Organizational Change Management, 30*(2):217–231.

Karmasin, M. (2002). Towards a meta ethics of culture-halfway to a theory of Metanorms. *Journal of Business Ethics, 39/2002*, 337–346. Kluwer.

Karmasin, M., & Apfelthaler, G. (2017). Integrated corporate social responsibility communication: A global and cross-CulturalPerspective. In S. Diehl, M. Karmasin, B. Mueller, R. Terlutter, & F. Weder (Eds.), *Handbook of integrated CSR communication* (pp. 237–251). Springer.

Karmasin, M., & Bichler, K. (2017). Corporate social responsibility in the media industry: Setting the benchmark or falling behind? In K. D. Altmeppen, C. A. Hollifield, & J. van Loon (Eds.), *Value-oriented media management: Decision making between profit and responsibility* (pp. 135–147). Springer.

Karmasin, M., Kraus, D., Kaltenbrunner, A., & Bichler, K. (2014). Media ethics as institutional ethics—the potential of corporate social responsibility. In S. Fengler, T. Eberwein, G. Mazzoleni, C. Porlezza, & St. Ruß-Mohl (Eds.), *Journalists and media accountability. An international study of news people in the digital age* (pp. 227–245). Peter Lang.

Kim, M., White, C., & Kim, C. (2019). Examining relationships among cultural factors and expectations of CSR. *Journal of Communication Management, 23*(4), 427–443. https://doi.org/10.1108/JCOM-01-2019-0009

Kim, Y., & Kim, S.-Y. (2010). The influence of cultural values on perceptions of corporate social responsibility: Application of Hofstede's dimensions to Korean public relations practitioners. *Journal of Business Ethics, 91*(4), 485–500.

Koinig, I., Weder, F., Diehl, S., & Karmasin, M. (2019). Media accountability with a special focus on CSR and the media industry: A topic of relevance? In T. Eberwein, S. Fengler, & M. Karmasin (Eds.), *Media accountability in the era of post-truth politics: European challenges and perspectives: European challenges and perspectives* (ECREA-series Routledge studies in European Communication Research and Education) (pp. 162–179). Routlegde.

Krainer, L. (2002). Medienethik als angewandte Ethik: Zur Organisation ethischer Entscheidungsprozesse. In M. Karmasin (Ed.), *Medien und Ethik* (pp. 156–174). Reclam.
Lai, C. H., & Fu, J. S. (2020). Humanitarian relief and development organizations' stakeholder targeting communication on social media and beyond. *Voluntas: International Journal of Voluntary and Nonprofit Organizations, 32*, 120–135.
Ligeti, G., & Oravecz, A. (2008). CSR communication of corporate Enterprises in Hungary. *Journal of Business Ethics, 84*, 137–149.
Liu, X. Y., Garcia, P., & Vredenburg, H. (2014). CSR adoption strategies of Chinese state oil companies: Effects of global competition and cooperation. *Social Responsibility Journal, 10*(1), 38–52.
Lock, I. (2019). Explicating communicative organization-stakeholder relationships in the digital age: A systematic review and research agenda. *Public Relations Review*. https://doi.org/10.1016/j.pubrev.2019.101829
Logsdon, J. M., Thomas, D. E., & Van Buren, H. J. (2006). Corporate social responsibility in large Mexican firms. *Journal of Corporate Citizenship, 2006*(21), 51–60.
Luethge, D., & Han, H. G. H. (2012). Assessing corporate social and financial performance in China. *Social Responsibility Journal, 8*(3), 389–403.
Maignan, L., & Ralston, D. A. (2002). Corporate social responsibility in Europe and the U.S.: Insights from businesses' self-presentations. *Journal of International Business Studies, 33*(3), 497–514.
Manetti, G., Bellucci, M., & Bagnoli, L. (2017). Stakeholder engagement and public information through social media: A study of Canadian and American public transportation agencies. *The American Review of Public Administration, 47*(8), 991–1009.
Men, L. R., Tsai, W. S., Chen, Z. F., & Ji, Y. G. (2018). Social presence and digital dialogic communication: Engagement lessons from top social CEOs. *Journal of Public Relations Research, 30*(3), 83–99.
Meyskens, M., & Paul, K. (2010). The evolution of corporate social reporting practices in Mexico. *Journal of Business Ethics, 91*, 211–227.
Moreno, Á., Navarro, C., Tench, R., & Zerfass, A. (2015). Does social media usage matter? An analysis of online practices and digital media perceptions of communication practitioners in Europe. *Public Relations Review, 41*(2), 242–253.
Morsing, M., Schultz, M., & Nielsen, K. U. (2008). The 'Catch 22'of communicating CSR: Findings from a Danish study. *Journal of Marketing Communications, 14*(2), 97–111.
Navarro, C., Moreno, Á., & Zerfass, A. (2017). Mastering the dialogic tools. Social media use and perceptions of public relations practitioners in Latin America. *Journal of Communication Management, 22*(1), 28–45.
Painter-Morland, M., & Deslandes, G. (2017). Reconceptualizing CSR in the media industry as relational accountability. *Journal of Business Ethics, 143*, 665–679.
Planken, B., Nickerson, C., & Sahu, S. (2013). CSR across the globe: Dutch and Indian consumers' responses to CSR. *International Journal of Organizational Analysis, 21*(3), 357–372.
Rim, H., & Dong, C. (2018). Trust and distrust in society and public perception of CSR: A cross-cultural study. *Social Responsibility Journal, 14*(1), 1–19.
Saxton, G. D., & Waters, R. D. (2014). What do stakeholders like on Facebook? Examining public reactions to nonprofit organizations' informational, promotional, and community-building messages. *Journal of Public Relations Research, 26*(3), 280–299.
Schlegelmilch, B., & Pollach, I. (2005). The perils and opportunities of communicating corporate ethics. *Journal of Marketing Management, 21*, 267–290.
Schmeltz, L. (2014). Introducing value-based framing as a strategy for communicating CSR. *Social Responsibility Journal, 10*(1), 184–206.
Soboleva, I. (2007). Corporate social responsibility in Russia: Peculiarities and problems. https://doi.org/10.5840/icr2007317.

Stahl, G. K., Miska, C., Puffer, S. M., & McCarthy, D. J. (2016). *Responsible global leadership in emerging markets, Advances in global leadership (advances in global leadership, vol 9)*. Emerald Group Publishing Limited, (pp. 79–106).

Szőcs, I., & Miska, C. (2020). Ethics in the context of cross-cultural management. In B. Szkudlarek, L. Romani, D. V. Caprar, & J. S. Osland (Eds.), *The SAGE handbook of contemporary cross-cultural management*. SAGE.

Takano, K. (2013). Communicating CSR: Case study of Japanese food industry. *International Journal of Business and Management, 8*(9), 111–121.

Tench, R., Bowd, R., & Jones, B. (2007). Perceptions and perspectives: Corporate social responsibility and the media. *Journal of Communication Management, 11*(4), 348–370.

Tingyang, Z. (2020). *Alles unter dem Himmel: Vergangenheit und Zukunft der Weltordnung*. Suhrkamp Verlag.

Toker, H. (2013). Media CSR forum. In S. O. Idowu, N. Capaldi, L. Zu, & A. D. Gupta (Eds.), *Encyclopedia of corporate social responsibility* (pp. 1643–1650). Springer.

Trommershausen, A., & Karmasin, M. (2020). Corporate social responsibility in Medienunternehmen. In J. Krone & T. Pellegrini (Eds.), *Handbuch Medienökonomie* (pp. 1235–1296). Springer.

Voci, D., Karmasin, M., Przybylski, P., Altmeppen, K. D., Möller, J., & von Rimscha, M. B. (2019). What is a media company today? Rethinking theoretical and empirical definitions. SCM (Studies in Communication and Media). (pp. 29–52).

Wang, Y., & Yang, Y. (2020). Dialogic communication on social media: How organizations use Twitter to build dialogic relationships with their publics. *Computers in Human Behavior, 104*, 106–183.

Watkins, B. A. (2017). Experimenting with dialogue on Twitter: An examination of the influence of the dialogic principles on engagement, interaction, and attitude. *Public Relations Review, 43*, 163–171.

Weder, F., & Karmasin, M. (2017). Communicating responsibility: Responsible communication. In S. Diehl, M. Karmasin, B. Mueller, R. Terlutter, & F. Weder (Eds.), *Handbook of integrated CSR communication* (pp. 71–89). Springer.

Zhao, T. (2020). *Alles unter dem Himmel: Vergangenheit und Zukunft der Weltordnung*. Suhrkamp Verlag.

# Sustainability as Cultural Practice and Media as Institutions of Change

**Franzisca Weder**

## 1 Introduction

Adaption to a changing climate and finding pathways to social, economic, and ecological sustainability is probably the biggest challenge of the twenty-first century. Academics, policymakers, CEOs, and campaigners increasingly recognize the contribution of culture for people to understand the world, to make sense of the world. Cultural sustainability or "culture in, for and as sustainability" is being explored and discussed in various thematic contexts, representing a wider range of practices and processes, like lifestyles, landscapes, artwork and artistic practices, heritage, tourism, and more (Soini & Dessein, 2016). However, there are two current challenges that lead to the chapter at hand: Firstly, facing a series of global crises (climate change, pandemic), and caused by social, political, cultural, and environmental impacts of economic and technological globalization processes, we are under risk (Beck, 2006) and live in a massively disrupted "threat society". Covid 19, countries boycotting the Paris Agreement (USA, 2020) and future COP conferences (Australia, 2021), natural hazards, ethnic terror, and war mark severe obstacles for the realization of the Sustainable Development Goals, outlined by the UN for 2030 (UN, 2021), and show their vagueness. In this situation, media play a key role in our society, not only as a source of information or "watchdog," but much more as a facilitator of public sense- and meaning-making processes. Media don't just offer us a window to the world, they don't present reality—they represent it. Thus, media of all kind, film, radio, blogs and podcasts, newspapers and magazines, and social media have the potential to define what sustainability means. They represent and possibly reflect on the public impression of the changing climate and stimulate conversations and negotiations *around* sustainability.

F. Weder (✉)
School of Communication and Arts, University of Queensland, Brisbane, QLD, Australia
e-mail: f.weder@uq.edu.au

 
F. Weder et al. (eds.), *CSR Communication in the Media*, CSR, Sustainability, Ethics & Governance, https://doi.org/10.1007/978-3-031-18976-0_3

While there is more and more research focusing on sustainability-related issues represented *in* the media (predominantly climate change, fracking, renewable energy, inequality issues, or business models like circular economy), there is a lack of literature revisiting original concepts of the role of the media in redefining and conceptualizing the idea of a culture of sustainability from a communication studies perspective.

Thus, in the chapter at hand, firstly, the existing literature on media business and CSR, sustainability communication, and Public Value is briefly presented. Secondly, the idea of media as institutions of change and sustainability as cultural practice defined and facilitated in the media is developed, followed by a discussion and outlook for future research potential.

## 2 Media, CSR, and Sustainability

In media and communication studies, media are perceived either as "channel" (film studies, newspaper- or social media-research, etc.) or as organizations (media economics, media politics, media ethics, etc.). From an organizational perspective, media follow economic principles, they have employees and stakeholders and produce a—rather specific—good, product, or service. As organization, they are embedded in the society, they are communicatively related to their audience(s). This embeddedness creates responsibilities, which are allocated to a medium, a radio station, or even a journalist or chief editor. The concept of business or "corporate responsibility" toward the society is mostly captured as "Corporate Social Responsibility" (CSR). Due to their very specific product (article, show, blog but as well: awareness, information, social orientation, etc.), media institutions or corporations deal with a dual or "twin responsibility" (Karmasin & Weder, 2008).

Thus, firstly, media corporates are a business, as such, they are responsible for their economic success, their impact on the environment as well as their "social footprint," including their relationship to employees and external stakeholders and audiences. The social impact (SI) on the society and the key audiences or publics as well as the need to get their license to operate (SLO) (Hurst et al., 2020) is discussed in various fields of predominantly CSR communication research (Rasche et al., 2017; Morsing & Spence, 2019; Diehl et al., 2017; Allen, 2016; Saffer et al., 2019). Even though, it has to be stated that the media have not (yet) been challenged enough to consistently introduce responsible business models and respond to the sustainability principles.

Secondly, next to the media's role and responsibility from a business perspective, they produce culture—or at least, cultural goods (issues, opinion, reputation, reality, social orientation, etc.) (Karmasin & Weder, 2008) and do have a responsibility for these products. Due to the "public value" that the product or "good" (potentially) carries, there is a second dimension of responsibility that needs to be considered, the so-called communicative responsibility (Karmasin & Weder, 2008) means the responsibility for *what* and *how* a medium communicates. To further understand communicative responsibility next to economic, social, and environmental responsibility *of* the media, the concept of "public service" (Murschetz, 2021) needs to be bridged with the existing CSR (Rasche et al., 2017) and media economics (Albarran et al., 2018) literature. Furthermore, sustainability as overarching normative framework and guiding principle for responsible behavior and action works as a key for bringing these concepts together and developing a framework for what makes media valuable to society.

Sustainability itself is a communication phenomenon of the twenty-first century. It started as an economic description of the use of resources with avoiding their depletion in order to maintain an ecological balance (WCED, 1987, p. 51); however, the concept of sustainability changed at the end of the 1990s into a future vision of growth, a vision of the future within the existing system of capitalism. With a massive wave of corporate communication, sustainability gained in fame and, above all, increased its public "glamor," it even developed into a master frame (Weder, 2021), used (and abused) in many contexts of meaning. And even media corporates like the Austrian Broadcasting Corporation (ORF) started to describe themselves as a company with social responsibility that shows social commitment and accordingly contributes to "sustainable development" (zukunft.orf.at, 2021).

However, from a public or society perspective, looking at the potential for media to contribute to societal change and transformation processes by developing new ways to communicate *for* change (Weder, 2021), we need to go beyond statements and communication of sustainability-related activities. Therefore, in the following, we argue, that today, at least in these times of a climate crisis and a global pandemic, communication is the key to learning and shaping the future and developing a culture of sustainability (Weder, 2021). Furthermore, we will develop a concept of sustainability as cultural practice and identify the media's role in facilitating learning, sense-making and change and their potential to create a culture of sustainability.

While there is a growing debate in academia and practice on concepts and strategies for communication *of* and *about* sustainability (Newig et al. 2013; Golob et al., 2013; Fischer et al., 2016; Allen, 2016), the idea of communication *for* change, for sustainable development is rather underdeveloped. First attempts are made by communication scholars (Weder, 2021) or in the areas of sustainable consumption (Newig et al., 2013) or education for sustainability (Weder & Milstein, 2021). Bringing in the idea of creation of a public value and the concept and idea of "public service," media organizations and institutions need to be re-conceptualized as playing a key role in the development of a culture of sustainability and in social transformation processes.

## 3 Culture of Sustainability *In* and *Through* the Media

In the previous sections of the chapter, sustainability as principle was presented as still being complex, and an institutionalization of sustainable development as still not being very well developed. By the same time, from a communication perspective, we observe a public "sustainababble," a cacophony of statements, frames, and narratives mainly driven by corporates and political institutions, which refer to the principle of restoration, regeneration, and transformation, but do not necessarily define or further negotiate it; the WHY and the WHAT precisely are mostly not further elaborated. As well, existing (mainly corporate) sustainability communication does not include the conflicts, abundance, or inconvenient truths, coming along with sustainable development and social change. True sustainability means to think more complex and to make more complex decisions. Therefore, it needs more communication, more storytelling, more communication that talks about conflicts, decision-making processes, dissonances, more space and time for these negotiation of what sustainability means, more communication to create public impressions of what sustainability is. And here media come into play.

If sustainability is conceptualized as principle of action and thus, as "culture," we change the perspective. In this case, culture is more than museums, theaters, and artwork; if our understanding of culture includes human–nature relationships and thus, or eco-cultural identity (Milstein & Castro-Sotomayor, 2020), then every communication process becomes part of this "culture." Culture as social practice (Reckwitz, 2002) includes activities which are a combination of knowledge and communication, which follow routines and patterns and, thus, which are repetitive. At the core of this idea is the concept of performativity, which means that with related individual action the collective and, thus, the culture itself can be reproduced (which means maintained or changed, modified). And from a communication and media studies perspective, media are not only a structural complex, a subsystem of the society; much more, understanding communication as social practice, media are "doing culture" (Hörning & Reuter, 2004, S. 10; Weder, 2021)—they represent culture as *text* and as *action*. Media products are culture, by the same time all related interactions and practices are culture, and represent culture. Thus, if media communicate more about sustainable practices, the principle of sustainability becomes part of our "culture," and social practices like consumption can change accordingly.

To summarize: just because there are 17 Sustainable Development Goals or just because sustainability is increasingly mentioned or used as master frame in policy documents, individual behavior and decision-making will not be more sustainable. However, sustainability can act as moral compass (Weder et al., 2020), when individual actors relate themselves to others, orientate themselves to another. If individuals in a certain cultural context act very similar, then there is the potential to change cultural practices with changing the context. Thus, from changed practices a new culture of sustainability emerges—via communication, via talking and communicating about it and thus, via establishing new patterns, routines, mentalities, and norms—by which the complexity of sustainability is reduced alongside. And if an individual started to act more sustainable in one area of his life, then this can be transferred to another.

With the media opening up space and scope for more sustainability communication *with* and *in* their products and services, there will be a "chain reaction" (Weder, 2021). Media link individual action, and "normalize" sustainable behavior. Thus, sustainability as cultural (media) practice then means:

- Creation of a sustainability discourse (SPACE)
- Sense-making: Negotiation of Meaning making (SCOPE)

If sustainability is perceived as "cultural practice," manifested in "text" (media products), then this includes intangible components such as beliefs, languages, and ways of life, as well as tangible elements, such as arts, or media reports, a new supplements on climate (Guardian, UK; Falter, Austria). They open up the space for further discussions and negotiations of meaning, they contribute to defining the contemporary notion of sustainable development and, thus, they take part in a cultural shift.

Therefore, media are responsible not only for their CSR strategy and communicating *about* it—but much more for offering more space and scope for communication *for* sustainability. Then, they facilitate change by normalizing sustainable practices. Here, we are looking specifically at Public Service Media (PSM), offering a "public good." They should know how to action a "cultural practice" (Weder, 2021) more than others do. Public Service Media are meant to create social discourses, create meaning, and support the formation of opinions. Thus, a culture of sustainability can potentially emerge through and in public discourse(s), particularly facilitated by PSM, understood as cultural institutions. They provide interpretations (frames) related to individual actions and decision-making conflicts (storytelling) and thus enable individual action to become regular and more "normal" and thus solidify as a type and practice. Accordingly, cultural work within the media includes defining *what sustainability actually means* (the WHY). Thus, Corporate Media Responsibility means that media as cultural institutions have the responsibility to create space and scope for action and communication for sustainable action and to make sustainable action a practice that then grows into a culture of sustainability—at least starting with the practices of journalism.

## 4 Outlook and Future Research Potential

Existing literature on Corporate Social Responsibility and communication or the media does usually not include a fourth dimension of sustainability—the cultural dimension. Not only from a sustainability perspective, but even from a communication and media perspective, this marks a gap in the existing CSR and media (management) research. Not only responsibility, but culture also has a "twin character." As described at the beginning of the chapter, media are responsible in an economic dimension and have additionally a communicative responsibility for what and how they communicate. Similarly, culture is the framework or context and condition for certain processes and communicative actions, but by the same time, culture is the communication itself. Culture is the rules of the game that we call society, so culture determines the rules of the individual games that we play (what we perceive as normal, how we perceive people of the other sex, how we treat our parents/vice versa).

Culture leans on communication, because all the rules of our social life and togetherness are determined and guided through (public) discourses. Therefore, we need more public discourses on, about and thus, for sustainability. This is the core responsibility of the media. This offers future research potential for media and communication scholars, cultural studies, media economics, and social sciences in general. Communication science and the humanities are essential to strengthening the feasibility, acceptability, and normalization of new principles like sustainability. We recommend more research using conversational approaches to understand how sustainability is presented in the media so far and the logic behind, talking to journalist, media managers, and new merging communicators like influencers, bloggers, etc. As well, language studies and discourse analyses in particular have the potential to better understand the status quo of negotiating and defining sustainability.

## References

Albarran, A. B., Mierzejewska, B., & Jung, J. (Eds.). (2018). *Handbook of media management and economics*. Routledge.

Allen, M. (2016). Strategic communication for sustainable organizations. In *Theory and practice*. University of Arkansas.

Beck, U. (2006). Living in the world risk society: A Hobhouse Memorial Public Lecture given on Wednesday 15 February 2006 at the London School of Economics. *Economy and Society, 35* (3), 329–345.

Diehl, S., Karmasin, M., Mueller, B., Terlutter, R., & Weder, F. (Eds.). (2017). *Handbook of integrated CSR communication*. Springer.

Fischer, D., Lüdecke, G., Godemann, J., Michelsen, G., Newig, J., Rieckmann, M., & Schulz, D. (2016). Sustainability communication. In *Sustainability science* (pp. 139–148). Springer.

Golob, U., Podnar, K., Elving, W. J., Nielsen, A. E., Thomsen, C., & Schultz, F. (2013). CSR communication: quo vadis? *Corporate Communications: An International Journal.*

Hörning, K. H., & Reuter, J. (2004). Doing culture. In K. H. Hörning & J. Reuter (Eds.), *Culture as practice* (pp. 9–15).

Hurst, B., Johnston, K. A., & Lane, A. B. (2020). Engaging for a social licence to operate (SLO). *Public Relations Review, 46*(4), 101931.

Karmasin, M., & Weder, F. (2008). *Organizational communication and CSR: New challenges for communication management and PR*. Lit-Verlag.

Milstein, T., & Castro-Sotomayor, J. (Eds.). (2020). *Routledge handbook of ecocultural identity*. Routledge.

Morsing, M., & Spence, L. J. (2019). Corporate social responsibility (CSR) communication and small and medium sized enterprises: The governmentality dilemma of explicit and implicit CSR communication. *Human Relations, 72*(12), 1920–1947.

Murschetz, P. C. (2021). Big data and public service media: A literature review of key challenges and some theoretical propositions pertaining to the context of digital transformation. *Medienjournal-Zeitschrift für Medien-und Kommunkiationsforschung, 44*(3), 69–86.

Newig, J., Schulz, D., Fischer, D., Hetze, K., Laws, N., Lüdecke, G., & Rieckmann, M. (2013). Communication regarding sustainability: Conceptual perspectives and exploration of societal subsystems. *Sustainability, 5*(7), 2976–2990.

Rasche, A., Morsing, M., & Moon, J. (Eds.). (2017). *Corporate social responsibility: Strategy, communication, governance*. Cambridge University Press.

Reckwitz, A. (2002). Toward a theory of social practices: A development in culturalist theorizing. *European Journal of Social Theory, 5*(2), 243–263.
Saffer, A. J., Yang, A., Morehouse, J., & Qu, Y. (2019). It takes a village: A social network approach to NGOs' international public engagement. *American Behavioral Scientist, 63*(12), 1708–1727.
Soini, K., & Dessein, J. (2016). Culture-sustainability relation: Towards a conceptual framework. *Sustainability, 8*(2), 167.
WCED—World Commission on Environment & Development. (1987). p. 51.
Weder, F. (2021). Sustainability as master frame of the future? Potency and limits of sustainability as normative framework in corporate, political and NGO communication. *The Sustainability Communication Reader: A Reflective Compendium*, 103–119.
Weder, F., & Milstein, T. (2021). Revolutionaries needed! Environmental communication as a transformative discipline. In *The handbook of international trends in environmental communication* (pp. 407–419). Routledge.
Weder, F., Tungarat, A., & Lemke, S. (2020). Sustainability as cognitive "Friction": A narrative approach to understand the moral dissonance of sustainability and harmonization strategies. *Frontiers in Communication, 5*, 8.
Weder, F., Karmasin, M., Krainer, L., & Voci, D. (2021). Sustainability communication as critical perspective in media and communication studies—an introduction. *The Sustainability Communication Reader: A Reflective Compendium*, 1–12.

# The Should, the Could, and the Would: On Paradox Constellations of CSR in the Media Business—and a Call to Action

**Lars Rademacher**

## 1 Introduction

As explained throughout this volume, media companies have a double responsibility as producer of news media and—at the same time—as classic corporation that is confronted with expectations from society to be a proper corporate citizen or speculations about greenwashing or misconduct—especially with respect to public service media. But when it comes to sustainability and CSR, media outlets face the challenge to explain what exactly their range of responsibilities looks like. We see a number or organizations who identify their main responsibility as reliable and accountable news producers whereas others look more at the context to fields like energy supply or environmental effects caused by media production and search for sustainable alternatives (e.g., green production). And furthermore, news production could be seen as one of the core products of media companies. Can it—at the same time—also be at the heart of CSR activities; wouldn't we expect news media to be accurate, on the point and high quality right away without further account of emphatic explanations on wider responsibilities that could also be covered with being reliable and accountable?

## 2 Learning Change and Making Sense

It has become obvious that changing behavior is a major task if you look at it from a broader perspective: changing the mindset of a whole society, giving reasons for rethinking one's own behavior is probably the biggest challenge when the "How?"

L. Rademacher (✉)
Department Media, Darmstadt University of Applied Sciences, Darmstadt, Germany
e-mail: lars.rademacher@h-da.de

 
F. Weder et al. (eds.), *CSR Communication in the Media*, CSR, Sustainability, Ethics & Governance, https://doi.org/10.1007/978-3-031-18976-0_4

of reaching UN Sustainability Development Goals (SDGs) is being discussed. And from a meta-perspective the whole discourse on transformation that we have experienced in the past years can be perceived as a new narrative to motivate society and create new reasons to follow a path to change. Underneath this meta-discourse, we find a wide range of methods and process models that have successfully explained to go about change on a more applied yet mediating level. Here, I would like to focus on the so-called Theory of Change that is often described as the "missing middle" between abstract programs and political planning, on the one hand, and the down to earth realization of goals on-site (UNDG Guideline, 2020). Theory of Change as a process model starts with the desired long-term goals and works backward from there on looking at the necessary conditions (outcomes) that must be in place (and how these are related) for the goals to occur (Wendt, 2021). The framework then helps identifying which type of activity or intervention will lead to the outcomes that have been defined as preconditions to the long-term goals (CfToC, 2021). The huge benefit of this model is to uncover links between interventions and outcomes which leads to better and more realistic planning. Therefore, it is indeed the missing "middle ground" between the high-complexity explanations offered by academics and science and the often-heard short-cut argumentations that are produced and repeated in the media.

If we want to learn how to change, we will need tools like the Theory of Change to combine the methods and the means. But change is not only brought about by linking intervention, precondition, and long-term goals. We do need two more steps. One is aspiration, the other one is called the "ideology of the first step." Using aspiration means that we have to give people cues of what it would be like to live their life in a partly different environment. What would it look like, what would it feel like? Taking up these cues and combining them with the cues that people produce themselves, it is possible to help them develop a feeling what this new normal would look like, would actually feel like. And it is essential to prevent retention and to bridge the gap between cognitively knowing which kind of change would be necessary and an openness to actually try it. Trying change, taking the first step, feeling into this different version of reality and shaping it by changing yourself, this is all the ideology of the first step is about: you should know what to do first, how to enter change. It is not a massive change but a first step, a small step maybe. If this works it generates trust, if this trust emerges a next step is possible.

We have to know what it should look like (Visions & Goals), we have to fill the gaps between interventions, preconditions, and long-term goals. And we have to invite people to first steps and lead them to slowly growing trust in change. As much as I admire the challenge the Extinction Rebellion and Stop Oil present so mature economies worldwide: if you want to reach the societal shift in large scale, there is no other way than learning to change and making sense—which means to envision a different future and to create the desire of being a part of it.

## 3 The New Role of Media Organizations

Knowing that this is the pathway to change, what is the specific role of media in this constellation? On the one hand, see that media have to represent reality, they have to present "what is." And—as Horst Pöttker (1999) once wrote—news journalism should "print and broadcast," this and only this would be his main obligation. But is that actually enough? Shouldn't journalism also present alternatives to the present to lay ground for a development in the future? Shouldn't journalism also test concepts and ideas? Forster new developments, create a public "maker's space" for new ideas for technological, social, and ecological improvement and innovation? And if he should, what is the specific role of Public Service Media (PSM) herein?

As Jens Müller (2022 in this volume) has emphasized, PSM were among the first to point at the climate crisis 30 years ago. But we have to ask back: has this brought about any change so far. Yes, we read more, we hear more, and we know more about the climate crisis and necessary adjustments. But did it have any impact on behavioral change? Or cause a change of mind?—Rarely. This is a reason why solution and constructive journalism has become very popular in recent years. After a surge of academic article and examples we now see a phase of consolidation (Lough & McIntyre, 2021) and theoretical development. Authors like Torsten Schäfer (2022) expect an intensified theory-practice-discourse that goes beyond classic academic reflection: Academia should also commit themselves to ethical standards and accept planetary boundaries. Sustainable development, he argues, has developed into a universal value like human rights, etc. Therefore, there is no question if climate crisis should be a focus part in news coverage or if sustainable development should be an ethical backbone of media organizations.

News media, according to Schäfer (2022), have the chance to come up with new narrations and narratives on climate issues. He calls "contextualizing" an art that is at the core of journalistic techniques. And this is a development that has potential also for local and regional journalism where climate change is also a local experience that can be linked to wider contexts. But nation-wide media, PSM especially, have a signaling role on this path to more sustainability.

Finally, I would like to unfold one last and—in my view—central argument why media have no other choice than to transcend the "would" and reach the "could": the relevance of traditional news media has shifted—not unidirectional, we have seen shifts in many directions. But the overall decline in circulation is a fact. And digital news products do not close the gap yet. On the other hand, there are multiple news sources that are biased by industry interests or political groups, radicals, climate change deniers and others. And in the pandemic, we have also seen this diverse scenario: a huge part of the public turning back to quality media while others were disappointed from what they believe was mainstreaming and suppression of opposing arguments. But over all this situation of crisis has proven the viability and the value of independent quality news media. This role cannot be filled by anyone else.

Media organizations giving orientation in a society under pressure, a society that is in the middle of a large-scale crisis, it is the central duty of responsible news journalism to open ways of thinking, to bridge the gap between knowing what to change and feeling and experiencing the change. And to invite the audience to try the first step by giving example. This might be a way to meet the often discussed goal #18: being sustainable as a media company.

## References

Center for the Theory of Change (CfToC). (2021). What is theory of change. https://www.theoryofchange.org/what-is-theory-of-change/ 10 August 2022.

Lough, K., & McIntyre, K. (2021). A systematic review of constructive and solutions journalism research. *Journalism*. https://doi.org/10.1177/14648849211044559

Pöttker, H. (1999). Öffentlichkeit als gesellschaftlicher Auftrag. Zum Verhältnis von Berufsethos und universaler Moral im Journalismus. In R. Funiok et al. (Eds.), *Medienethik—die Frage nach der Verantwortung* (pp. 215–232). Bundeszentrale für politische Bildung.

Schäfer, T. (2022). Transformation und Wiederverortung. Herausforderungen angesichts Klimanotstand und Artensterben. *Communicatio Socialis, 55*(2), 160–175.

United Nations Development Group (UNDG). (2020). Theory of change. UNDAF companion guidelines. https://unsdg.un.org/sites/default/files/UNDG-UNDAF-Companion-Pieces-7-Theory-of-Change.pdf.

Wendt, K. (Ed.). (2021). *Theories of change. Change leadership tools, models and applications for investing in sustainable development*. Springer.

# Public Value for Public Service Media? A Case Study Analysis of Austria's ORF

**Paul Clemens Murschetz, Eduard Frantz, and Niko Alm**

## 1 Introduction

Public Service Media (PSM), that is mainly "legacy" public broadcasting, of which the British Broadcasting Corporation (BBC) is the best-known example, are often established by law. They are non-partisan by design, independent from government and offer their services for the benefit of society at large.

Their legal remit is to inform, entertain, and educate the public who pay for them. But in doing so they must also protect their independence and ensure accountability. Their core mission is to serve the public interest.

There are plenty of examples around the world where governments are financially supporting media in the public interest—ideally without wielding editorial influence. Particularly in democratic countries, government intervention to preserve and promote "public value" dimensions of news and quality journalism is structurally formative.

In Europe, PSM are represented by the European Broadcasting Union (EBU). The world's largest alliance counts 69 members in 56 countries in Europe and beyond (www.ebu.ch/about/members), most prominently the BBC (UK), ARD and ZDF (Germany), DR (Denmark), RAI (Italy), France Télévisions (France), YLE (Finland) in the field of TV broadcasting, but also a fleet of PSM radio stations,

P. C. Murschetz (✉)
Murschetz media consulting, Salzburg, Austria
e-mail: paul.murschetz@gmx.net

E. Frantz
HHL Leipzig Graduate School of Management, Leipzig, Germany
e-mail: eduard.frantz@hhl.de

N. Alm
Average 2042 GmbH, Vienna, Austria
e-mail: niko@alm.net

 
F. Weder et al. (eds.), *CSR Communication in the Media*, CSR, Sustainability, Ethics & Governance, https://doi.org/10.1007/978-3-031-18976-0_5

for example, Deutschlandradio (Germany), an ad-free PSM radio station from Germany.

However, across Europe and beyond, PSM are increasingly under pressure. Digitization and emergent technologies are changing the use of media rapidly, with streaming services and online platforms gaining in importance and making it harder for legacy PSM to hold their ground.

This means that supporting the role of trustworthy, anti-partisan quality news journalism and news media in a democratic society—primarily taking the form of license-fee funding for public service broadcasting but also direct grants and indirect support to private printed media incumbents—seems ever more important.

Austria is no exception. The Austrian Broadcasting Corporation (Österreichischer Rundfunk, ORF) is the country's main PSM broadcaster (ORF, 2017a, 2017b, 2019, 2021). Even though private broadcasting companies were allowed only to enter the market in the late 1990s, the ORF has been offering its services since 1955 and up until today dominates broadcasting (and online) services provision in Austria.

It runs three nationwide (Ö1, Ö3, FM4) and nine regional radio channels—one for each province (Bundesland). Its TV portfolio includes two general interest channels (ORF 1 and ORF 2), one cultural-instructional channel (ORF III), one Eurovision-wide version of ORF 2 and a sports channel (ORF Sport +). ORF also takes part in the international special-interest channel 3sat.

This study examines public value theory and critically discusses its validity in explaining impacts on principles and remit of Public Service Media (PSM) in the digital era. The stakes are high as PSM such as the ORF need to both modernize and remain publicly responsible at the same time, for all audiences.

The purpose of this study is two-fold: First, it theorizes on the concept of "public value" and presents some dominant scholarly debates around the concept's approaches to measurement and funding as applied to the media industry.

By this, it argues that it is rather unclear how the public value mission is economically valuable for PSM and more importantly how, if at all, public value might be socially beneficial for media audiences and society at large. Hence, we principally ask whether public value for PSM is as strong and valid a concept to generate and prove both economic value and societal welfare.

And second, we look at how Austria's ORF, the country's dominant media and key PSM player, is coping with the current transformation of public service broadcasting through means of its public value "TransFORM" project. This implies gaining more insights to the concept of public value in the media from various selected facets.

We argue that the traditional concept of "public value" as widely scholarly agreed in PSM research is likely to effectively contribute to mastering the fundamental challenges of PSM, that is mostly one of digitalization, accountability, and openness for all citizens in a given country.

In our view, the traditional public value concept is elusive in character. It has high face validity, but only narrowly based explication, little consensus on meaning, and only recently emerging empirical verification.

To be truly effective, public value for PSM needs to be defined both more broadly and narrowly at the same time. More broadly in mission and vision, and more narrowly in proven impact. Evidence-based performance measurement of PSM remains vitally important.

There, processes of digitization have stepped up competition in the programming, advertising, and viewer markets of broadcasting, which turn out to endanger the continued existence of public service broadcasting in Austria.

A case study analysis of Austria's Public Service Broadcasting Corporation (ORF) thus reveals key challenges in delivering the promises of PSM's contribution to society in this process of transformation.

Given above challenges to media organizations, this study attempts to answer the following four questions:

- $RQ_1$: What is public value in theory?
- $RQ_2$: How can public value be measured?
- $RQ_3$: How do public value concepts best relate to PSM?
- $RQ_4$: What funding models are most effective for PSM?

## 2 Public Value Theory

"Public value" is a young concept and combines a mix of economics and management theory. It was introduced by Harvard University Professor Mark Moore in 1995 with the aim of laying out "a structure of practical reasoning to guide managers of public enterprises. It presents a general answer to the question of what public managers should think and do to exploit the particular circumstances they find themselves in to create public value" (p. 1). Moore described the challenge as changing the approach of public administrators who conventionally saw their duty as simply being to follow efficiently and effectively the legislative-based mandates that set out the purposes of particular public enterprises and came to conclude that the goal of operational activity in the public sector is the creation of public value. However, Moore also admitted that public value cannot be "measured" in principle.

The British BBC was the first broadcaster to engage with this new public value doctrine (Collins, 2007). It drew up its own manifesto (BBC, 2004), which on the one hand justified the levying of fees, but on the other hand also provided for accountability to the fee payers: The services provided were to be assessed in terms of their public value.

The academic debate about the public value of broadcasting across all offerings and age groups is complex and quite confusing (Alford & O'Flynn, 2009; Benington & Moore, 2009; Gundlach, 2011; Karmasin et al., 2011; Talbot, 2009). This is mainly due to the hybrid intellectual architecture of public value, which leads to competing scientific discourses.

Three schools are fighting for academic hegemony: First, the beginning of public value research dates to 1995, with Mark H. Moore's seminal book *Creating Public*

*Value: Strategic Management in Government* (1995). Then, Moore presented the public value concept as a normative theory of strategic management in the public sector and saw it as the equivalent of shareholder value in the management of private companies. Social value creation has since become a corporate duty.

In contrast to Moore's management perspective, Barry Bozeman's work is concerned with the societal level (Bryson et al., 2015). This is reflected in Bozeman's definition of public values as "those that provide a normative consensus about the rights, benefits and privileges to which citizens should (and should not) be entitled; the duties of citizens to society, the state and each other; and the principles on which governments and policies should be based" (Bozeman, 2007, p. 17).

Unlike Moore's (1995) and Bozeman's (2007, 2009) approaches to public value, Meynhardt (2015, 2019) offers a basis for a more refined conception of value. Not only does he anchor public value in a relational approach to value and embed it in an ontological basis in basic human needs, but he also relates it to a notion of the public rooted in individuals' representations and interpretations. Simply put, public value reflects basic human needs, and basic needs form the foundation for public value.

In essence, the academic debate about public value creation of the media is complex. It raises questions about the value of media content and the quality of services offered, particularly regarding the values to be conveyed for society. Further, it intrigues about PSM's legal obligation to produce societally relevant content, distribute over diverse platforms and channels, and accepted and consumed by heterogeneous audiences (Dobusch, 2018).

A recent study by Faulkner and Kaufman (2018) highlighted the different dimensions of Public Value measurement through an accurate and systematic analysis of the literature. Through a qualitative synthesis of the topics analyzed in the literature, the authors identified four analytical themes of public value measurement: (1) "outcome achievement"; (2) "trust and legitimacy"; (3) "service delivery quality"; and (4) "efficiency." These themes represent four key constructs that the reviewed studies suggest are important dimensions of public value measurement.

Ad (1) *Outcome achievement*: An outcome would be, for example, a social economic, ecological, or cultural outcome that is perceived as an improvement of the public sector, such as increased security and good medical care. Bozeman and colleagues (2015) mentioned several such outcomes, including national security, food safety, human health, environmental protection; and access to knowledge. Benington (2009, 2011) similarly suggests that benefits to economic activity and employment, social capital and cohesion, and environmental outcomes (e.g., reducing pollution, waste, and global warming) could be considered as aspects of public value. Similarly, several authors note that public value can be indicated by improvements in social outcomes (Brookes & Wiggan, 2009; Talbot & Wiggan, 2010; Meynhardt & Bartholomes, 2011; Conolly, 2013; Bracci et al., 2014; Spano, 2014; Moore, 2013). When referred to the media industry, quality media may achieve such outcome through the so-called positive externalities in the consumption of, for example, regional news, cultural programs, or the broadcasting of parliamentary debates.

Ad (2) *Trust and Legitimacy* refers to the extent to which an organization and its activities are trusted and perceived to be legitimate by the public and by key stakeholders. This dimension includes the extent to which the public trust the particular institution (Talbot & Wiggan, 2010), trust the programs or services delivered by institution (Heeks, 2008), and perceive an institution to be delivering services transparently and fairly (Meynhardt & Bartholomes, 2011). Talbot and Wiggan (2010) argue that trust and legitimacy are the heart of public value creation, largely because organizations that are trusted and perceived to be legitimate are best able to secure the support needed to carry out their objectives.

Ad (3) *Quality of the service* refers to the extent to which services are experienced as being delivered in high-quality manner that is considerate of users' needs. It is expected to be maximized when individuals who interact with the service are satisfied, and when they perceive the services to be responsive to their needs, accessible, convenient, and incorporate sufficient citizen engagement (Benington, 2009, 2011; Spano, 2014). Although no studies explicitly developed or validated quantitative measures of service delivery quality, at least two studies provided suggestions that may be useful for developing such a measure. Meynhardt and Bartholmes' (2011) measure of public value included items related to perceived service quality, customer satisfaction, and responsiveness to feedback.

Ad (4) *Efficiency* refers to the extent to which an organization is achieving maximal benefits with minimal resources. It is expected to be high when the benefits provided by an organization are perceived to outweigh the costs of that organization (Talbot & Wiggan, 2010), when "unnecessary" bureaucracy is avoided (Meynhardt & Bartholomes, 2011), and when an organization is perceived to offer "value for money" (Collins, 2007; Talbot & Wiggan, 2010).

By and large, it is crucial that an organization's value contribution to society is also experienced and accepted by individual recipients. For PSM, such acceptance is signaled, for example, by a license fee payers' willingness to pay. We posit here that only through individual recipients' appreciation through paid-for consumption does social acceptance for the collective good of public service broadcasting arise. The UK's Work Foundation describes the process of individual appreciation as follows: "Public value is what the public values, and it is the role of public managers to help determine through the democratic processes of deliberation and public engagement what social outcomes are desirable. It is through such processes that public managers can help to articulate collective citizens' preferences and thereby redress the 'democratic deficit' between public services and citizens" (Horner et al., 2006, p. 7).

For public value not become an empty formula, a central problem thus must be solved: What counts as public value is not only a problem of definition and agreement upon particular interest groups in society in the sense that public organizations automatically create public value as a matter of their remit, but, more essentially, predominantly a problem of measurement and proven evidence in public value performance (Hills & Sullivan, 2006). When applied to media organizations—a topic which we tackle in more detail below—consumers of public value programs may derive benefits from PSM programs and services only if these become a proven and real value that satisfies their preferences.

If this individual media benefit is not generated, no willingness to pay and hence no social value is generated, and thus ultimately no public sphere created. In the media economy, individual consumer preferences determine the benefit of a market offering.

Both PSM and private commercial media have to manage this major challenge in meeting the mission of a public value organization in times of transformational competition.

## 3 Public Value Measurement

We are living through a period of immense disruption in the media industry. The creation of the internet and social media and all that it has wrought—networked interactivity, immediacy, and fragmented audiences—has set in motion the destruction of the old business models that supported traditional media companies in the past such as broadcast TV, commercial radio, and newspapers. And, worse yet, the industry is experiencing an economic crisis and "legacy" media see a painful and ongoing decline in revenues as paid display ads, subscriptions, and direct sales shrink, while costs are still on the rise.

PSM are greatly challenged too. They need to deliver more and better services despite unchanging or diminishing public resources and support. But how can public accountability and public value best be measured?

Traditionally, at its 2012 General Assembly, members of the European Broadcasting Union (EBU) from Europe and beyond unanimously adopted six key values which define why public service media (PSM) are indispensable in true democratic societies (EBU, 2012).

The following traditionally acclaimed core PSM values are also well resonated and agreed-upon as seminal within scholarly PSM research (Eberwein et al., 2019; Van den Bulck, 2015).

- *Accountability* (i.e., discourse with the audience, transparency to the outside world, fulfillment, and control of the public service)
- *Diversity* (i.e., pluralism regarding program genres, views, and staff)
- *Excellence* (i.e., professionality and high-quality standards to meet audience expectations)
- *Independence* (i.e., non-partisan services and content delivery independent of political, commercial, and other influences, committed only to the interests of the public, committed to democracy)
- *Innovation* (i.e., intensive promotion of new formats, new technologies, and new ways to connect with the audience)
- *Universality* (i.e., the creation of a forum for all members of the society, unrestricted access for all on all technical platforms)

Certainly, these core values must be operationalized and measured to become evident and real. To do this, key figures of PSM as value-driven media organizations must first be identified, which are then combined in a meaningful way in a so-called Public Value Scorecard (PVSC) model. PVSC can provide a valuable decision-making tool for the management of PSM (Cobbold et al., 2004).

In fact, the European Broadcasting Union (EBU) and the German ARD—among others—have developed scorecard frameworks to evaluate the social value of public service broadcasting (EBU, 2017, 2020a).

Besides the problems in properly and successfully managing industry change (e.g., of digitalization and globalization), all above-mentioned organizations produce contents that are not ordinary commercial goods but goods with a social impact. For example, when media companies provide information about public health, it can affect the way people think or behave; therefore, media managers must take into consideration their social power of shaping people's opinions and beliefs.

The core element of the PVSC model is its cause and effect logic. The PVSC links corporate strategy (Input) with the customer view (Output), this with the process logic (Process) and finally society level (Outcome). The public value of PSM is thus multi-dimensional. As such, it can thus become subject to testing various hypotheses based on relations between the variables involved.

For example, a recent study by the Austrian Academy of Sciences has raised the questions as to whether government financing of public service media would directly influence audience success. Hence, "broadcasters with higher public funding achieve higher market shares, greater relevance as a trustful source of information and higher levels of trust in independence from external influences" (Eberwein et al., 2019, p. 143).

Consequently, government-funded media would achieve higher levels of trust and acceptance by the public, and hence achieve higher chances to contribute to influencing public opinion and shape public discourse. However, this may backfire in a sense that government funding may capture the media through controlling media by not financing independent journalism but choosing to fund instead media outlets that advance the government agenda and the interests of its allies and supporters, either political groups or businesses.

The University of St. Gallen in Switzerland has developed a "Public Value Scorecard" (PVSC) to help organizations understand and increase their value to society. The questions raised by this model are: What value are large organizations bringing to society? And, when they are getting it right, how can and should corporations build trust and gain legitimacy within society?

Meynhardt et al. (2009, 2015, 2019) PVC model focuses directly on public value creation along five theory-based basic public value dimensions. These are:

- Task fulfillment: Does the organization fulfill its core task or "is it useful?"
- Economic viability: Does the organization use economic resources efficiently or "is it profitable?"
- Morality: Does the organization behave decently or "is it decent?"
- Quality of life: Does the organization contribute to quality of life or "is it a positive experience?"

- Social cohesion: Does the organization contribute to the common good or "is it welfare-enhancing?"

Meynhardt (2009) roots the notion of value in psychological needs theory. Seen this way, public value must primarily be measured subjectively, as an individual's perception of various beliefs, attitudes, values, sentiments, and feelings when consuming a specific PSM's service.

Value measurement must thus be linked to individuals' perceptions of, for instance, (1) the basic need for orientation and control, (2) the basic need for self-esteem maintenance and self-esteem enhancement, (3) the basic need for positive social relationships, and (4) the basic need for pleasure gain and displeasure avoidance (when following Seymour Epstein's Cognitive-experiential self-theory (2003) (Table 1).

**Table 1** The Public Value Scorecard (PVSC): Public value dimension, Objectives (general), and Objectives (PSM-related)

| Public value dimension | Objectives (general) | Objectives (PSM-related) |
|---|---|---|
| **Task fulfillment**<br>*Is it useful?* | • Fulfills public tasks per remit<br>• Is professionally recognized<br>• Is legitimate, trustworthy, and independent | • Fulfills its legal remit<br>• Is accountable, diverse, excellent, independent, innovative, and universal[a] |
| **Economic viability**<br>*Is it profitable?* | • Strives for utility, efficiency, and effectivity<br>• Encompasses subjectively perceived financial or economic value<br>• Contributes to economic welfare<br>• Provides good quality to people | • Performs well in its core business<br>• Is economically viable<br>• Is professionally recognized<br>• Contributes to economic welfare in the region |
| **Morality**<br>*Is it decent?* | • Based on need for self-worth and dignity<br>• Related to subjective moral and ethical standards<br>• Striving for equality, fairness, ethicality | • Behaves decently<br>• Behaves fair<br>• Acts ethically correct and just<br>• Respects the human dignity of individuals |
| **Quality of life**<br>*Is it a positive experience?* | • Based on individuals' need to avoid pain and maximize pleasure<br>• Ranges from basic need to survive to positive hedonistic experiences<br>• Strives for happiness, joy, relaxation, beauty | • Raises positive emotions<br>• Contributes to the quality of life<br>• Is enjoyable for people<br>• Is pleasant for people<br>• Contributes to the happiness of the people<br>• Contributes to the well-being of the people |
| **Social cohesion**<br>*Is it welfare enhancing?* | • Based on need for social relatedness and belonging<br>• Strives for positive relationships, social identity, or belonging<br>• Ranges from belonging to cooperation and solidarity | • Contributes to social cohesion<br>• Creates a community<br>• Has a positive effect on social relationships<br>• Contributes to solidarity |

Source: Adopted from Meynhardt and Jasinenko (2020)
[a] Questions related to PSM core value objectives following EBU (2012)

## 4 Austria's ORF: A Case Study Analysis

When looking to media more particularly, PSM value can no longer be measured solely based on traditional headline performance metrics successes such as audience share and reach.

Instead, today's PSM performance measurement needs to be based on alternative quality-based measures, e.g., audience satisfaction and approval ratings, or "value for money" measures. The World Economic Forum's latest report (2020) has put it as follows: "Media don't just help pass the time; they keep people informed. Increasingly, media create shared cultural moments and capture identity. We need a more dynamic measure to calculate the real value of media to society." (WEF, 2020). In gist, PSM need to rethink performance metrics and how public values may be viewed by the public and its members.

More recent studies emphasize the wider implications of public value creation in public media. Over the summer of 2020, against the backdrop of the COVID pandemic, Mazzucato et al. (2020) developed new metrics that can assess how the BBC delivers against its mission, creates public value, and co-shapes markets. In their report Creating and Measuring Dynamic Public Value at the BBC, the authors tackled questions of how to measure the BBC's public value more dynamically and holistically. By widening the picture beyond mere audience reach numbers, it identified three layers of public value in that the BBC creates broader value for individuals, industry, and society at large. In doing so it has created value for individuals in their experiences as consumers; for society in contributing to a more inclusive, tolerant, and diverse culture; and in industry in taking the kinds of risks necessary for new markets to emerge.

As proven, audience share and reach remain important. "Consumption figures help to show the BBC's reach, efficiency, and commercial success—these are key indicators of public value as without successful and widely viewed programming, the BBC would not have the legitimacy and scale to engage the wider public. But by widening the picture beyond numbers, we can see that the BBC creates broader value for individuals, industry, and society" (Mazzucato et al., 2020, p. 23; see also, Mazzucato, & Semieniuk, 2017; Mazzucato, 2016).

Further studies identified even a wider range of additional dimensions of public values that can be applied by PSM (Horner et al., 2006), such as public satisfaction, economic value (creation of economic activity/employment), social and cultural value (social capital/cohesion), political value (democratic dialogue, public participation), ecological value (sustainable development), service quality (reception, satisfaction, choice, fairness, cost), financial performance (income, expenditure value for money, efficiency), non-financial performance (customer satisfaction, service quality), social value from the user's perspective (tangible economic value from the administration's perspective, intangible economic value from the administration's perspective), trust and legitimacy, and the protection of civil rights.

Importantly, according to Mazzucato et al. (2020), there is a more basic need to return to an understanding of the state as a collective value creator, and not limit its role to stabilizing markets in crisis. By acting as investor, inventor, innovator, and consumer platform at the same time, the BBC would play an essential role in developing the UK's infrastructure for digital media innovation.

In what follows, we look to Austria as a best-practice case study to apply some selected issues of public value theory to the field of public service media. By using case study analysis technique, we show how the concept of Public Value is practically applied by Austria's PSM ORF. We start with the well-known proposition that public value creation and delivery is the very legitimacy of PSM. Of course, Austria is no exception. There, however, the public value debate was developing only late but finally gradually took hold.

In 2010, it resulted in some changes to the ORF Act (ORF-G). Since then, the Österreichischer Rundfunk ("Austrian Broadcasting Corporation," ORF)—the country's national public service broadcaster—has been cultivating a well-organized public value culture, business modeling, and reporting, based on its legal remit.

The central public value provision of the law is the so-called public service core remit, according to which an overall balanced program must contain demanding content on an equal footing. The annual and monthly schedules of television are to be drawn up in such a way that the ORF is obliged to broadcast demanding programs, at least in the main prime-time programs (from 8 to 10 p.m.).

The Act also states that the ORF needs to pay attention to the distinctiveness of the ORF in terms of content and appearance vis-à-vis commercial broadcasters. In addition, program and services quality must be cross-checked on an ongoing basis. Further, its broadcasts and services in the fields of information, culture, and science must be characterized by high quality.

## 4.1 Public Value Remit

The ORF is an independent PSM and defined as a foundation under public law. By its legal remit alone, it is committed to create public value. The ORF channels offer an extensive set of cross-genres programs in information, science, culture, sports, entertainment, as well as religion.

Every day, 49% of Austrians are watching ORF TV programs (3.7 million viewers) and these numbers rise to a weekly reach of 74.4% (5.6 million viewers). In 2018, the two main TV channels (ORF 1 and ORF 2) together reached 29.8% market share (people older than 3 years), which is the 8th highest figure among all public service media providers in Europe. The entire ORF television-fleet (also including ORF III and ORF Sport+) achieved a market share of 31.8% in 2019.

And, compared to commercial broadcasters, attention is paid to the distinctiveness of the ORF in content and services offerings.

Being a foundation under public law, it needs to fulfill several duties and mandates laid down in the ORF Act.

This includes investing all its revenues, i.e., radio and TV license fees (50%) and advertising and other revenues (50%), into its programs and services for its audiences.

Consequently, this is how it aims to fulfill its cultural and democratic functions. It provides Austrians with information, education, and entertainment, can influence the democratic polity, and ignites public discourse among various societally worthy topics that create a broader diversity of views, attitudes, and opinions among its audience.

But all this does not happen in vain. One current issue is whether the ORF will successfully master its transformation process from a public service broadcaster to a public service media platform. Back in 2020, ORF Director-General Alexander Wrabetz already presented "12 Theses on Austria as a Media Location." These were intended to "set out the framework conditions for a future-proof further development of Austria as a media location for the benefit of the audience." By this, he attempted nothing less than to envision future (survival) strategies of public service broadcasting in times of radical media change.

The core idea is: "For the future of Europe and European societies, maintaining an independent, diverse, transparent and sustainably viable media ecosystem is of central importance" (thesis 1), which is why "a strong ORF is indispensable for a strong Austrian media location" (thesis 6). The question is merely what definition of "strong" is appropriate in that context.

A solidarity is needed above all against the "dominance of German media groups in television," the "superiority of global players in the online sector" and "globally operating pay groups." The Director-General set the goal as follows: "The common goal of Austrian media companies must be to keep as much advertising money and added value as possible for the financing and production of media content in Austria," said Wrabetz (thesis 4), ignoring the political concept of a European Single Market.

The fact that the ORF chooses a problem-oriented approach to seek common solutions for the complex challenges ahead is good and right, even if in the same breath the ORF then immediately follows up and demands "digital development freedom in the sense of the audience" (thesis 8), i.e., finally wants to take its fate into its own hands.

To sum it up, ORF programs of public value should not only comply with the legal mandate but should more particularly trigger sentiments with the audience that its core values reflect to strengthen the Austrian society as such.

Here, it is worth recalling the ethos of PSM: "The value of a public service medium is demonstrated by the quality and diversity of the public space they create and shape with other important actors in society. A space where everyone can meet with everyone, where people can inform themselves, discuss, relax, enjoy, educate, sometimes argue, create ideas, listen. A space that basically everyone can help shape." (Ingrid Deltenre, 2012).

## 4.2 Public Value Reporting

Since 2015, the ORF publishes its own annual "Public Value Report," that is a performance and benefit balance to represent ORF activities and services (see ORF Public Value—Start).

Since then, five value dimensions ("individual," "social," "national," "international," and "corporate" value) and 18 performance categories keep record of ORF's media value, performance, and ensuing benefits for its audiences. The dimensions and categories are based on the ORF Act, ORF regulations, and the ORF code of conduct.

Each year's ORF Public Value Report typically focuses on one specific key issue, for example, "Opening Up" in 2015/6, a title to indicate ORF's and its employees' ambitions to reflect on citizen participation, quality journalism, and content and technology innovations in times when more open relationships between PSM and audiences were becoming ever more important.

The ORF's Director-General is responsible for the creation of this quality assurance system. The ORF Foundation Council (Stiftungsrat), which is politically staffed and votes in secret, must approve of it. The review of whether the quality criteria have been met is then to be carried out by an expert. This expert is also appointed by the Director-General with the approval of the Foundation Council.

The latest ORF Public Value Report 2021 ("TransFORM") deals with "101 Questions on the Digital Future of the ORF", based on the five quality dimensions: "Digital Me," "Digital Us," "Digital Local," "Digital Global," "Digital Platform". Data are enriched with reports by ORF employees, media experts and scientists on the topic of digital transformation.

Indeed, the big question is to safeguard the future viability of Austria's largest media provider. Disruptive developments in communication technologies have led to massive changes in media production and media use and have disrupted the media market in the hotly contested terrains of television, streaming, social media, and the internet. The ORF has received strong competition, while trust, especially of younger people, has faded.

In 2021, the ORF started the "TransFORM" process, a new change process initiated by the Public Value Department of the ORF. Building on prior research and analyses of challenges of the digital transformation on public value and the role of the PSM in the digital era (see, ORF Public Value), "TransFORM" is to:

- Analyze concrete opportunities of change by commissioning an international study
- Undertake the biggest survey ever on expectations of under 16- and 29-year-olds and their perceptions about the role of the ORF in times of change
- Invite experts and innovators to discuss the challenges in public ("ORF-DialogForum") to elicit a response and further dialogue on the change plans

Principally, with "TransFORM," the public broadcaster ORF seems to walk down the right path. It critically opens the debate over what constitutes public value for PSM in Austria in times of fundamental change.

"TransFORM" identified five quality dimensions and 18 performance categories which keep record of ORF's media performance, value, and benefits for the audience.

- *Individual Value*: ORF establishes a trustful[1] fundamental understanding of the world through extensive and factual accurate information. ORF service programs disseminate knowledge and support for numerous everyday life questions. ORF provides quality entertainment which means a versatile offer and quality standards, which consider requirements and high standard boundaries without advertising breaks. ORF takes responsibility and supports a direction toward common welfare
- *Social value*: ORF's cultural duty is to connect majorities with minorities. ORF news coverage offers orientation and support to understand the overwhelming news offer of the complex world. The foundation to this is to acknowledge the diversity of society. By providing the society a public platform close to the society's interests, demands, and worries, ORF can build strong and close relationships with the public, which is of great importance to ORF. Additionally, ORF has an integration purpose. ORF connects and brings people together regardless how diverse they might be
- *National value*: ORF regional coverage signifies living federalism. Encouraging the Austrian culture and creativity generates appreciation. ORF supports the discourse of Austrian identity with historic documentaries and initiatives
- *International value*: ORF anticipates itself as the bridge between the World, Europe, and Austria. Amongst other components this is achieved through the support of the biggest Austrian correspondent network and the assistance of European broadcasters and media networks
- *Corporate value*: Transparency and confirmability of the company as well as constant progressions of the employee's expertise are necessity to ORF. Through quality assurance ORF may implement what the audience demands: the capability of innovation in radio, TV and online

In any case, skepticism about meeting these value goals and principles prevails. Today, PSM are facing highly turbulent environments. The streaming services of YouTube, Netflix, and Amazon Prime are clearly ahead of linear television among young audiences.

Nevertheless, only with more intelligence, interaction and convergent infrastructures will the ORF succeed as a powerful PSM and become a powerful player in the "platform society" (Van Dijck et al., 2018). Why? Because ORF's claim to hegemony is crumbling in the convergent media world. There, innovation, variety, fragmentation, and diversity are structural principles. In short: The relevance of

[1] Items in italics are written in big letters in the report to signify its importance.

the journalistic competition in shaping public opinion is increasing regarding reach, displacement through other media, journalistic and political relevance, moving images and topicality. This weakens the ORF.

Consequently, it is necessary to rethink the role of PSM in a more fragmented society. On the one hand, this involves refining the societal contribution of public service. Starting from the notion that PSM should, as McQuail (2010, p. 178) put it, "serve the public interest by meeting the important communication needs of society and its citizens," these needs (e.g., contribution to democratic governance and culture, production of information and knowledge, cohesion and integration, or progress) and the ways PSM can address these needs in unique ways other media cannot have to be identified. On the other hand, it is also necessary to modernize the ways in which PSM provide their contribution to society. Beyond producing content for all kinds of distribution channels, platforms, and usage scenarios (ranging from the living room to mobile consumption), PSM have the chance to involve citizens in production and to evolve the ways in which their content reaches audiences (e.g., personalization-based services). Furthermore, it is necessary to discuss how the contribution of PSM to society can be measured.

## 5 Public Value Financing

Certainly, change implies risks and uncertainty. Hence, to safeguard funding sustainability for PSM remains pivotal.

Financial security is an essential building block for safeguarding PSM offerings. More strongly, for PSM to operate effectively, sound political, legal, and economic framework conditions are primarily important. This safeguards journalists to operate without interference or fear of violence, and PSM enjoy stable legal and business conditions that pay decent salaries, discourage media corruption, and promote sector-wide integrity.

However, and more generally, the PSM's legitimacy has come into question as its share of television consumption has markedly declined in the digital era.

Stressing the role and importance of public value offerings such as through EBU's public value projects is one way to mitigate this pressure on the license fee.

Another one is the Public Value Test (PVT) which typically assures that legal public value obligations are properly met, and the fees are closely tied back to the production and dissemination of high-quality content in the public interest (Donders & Moe, 2011).

Scholarly research has proven that a PVT is a solid and thorough evidence-based process which considers both the public value and market impact of proposals (Knoll, 2012).

For example, when PSM introduce new programs, shows, or technologies (such as the BBC's catch-up TV service of the iPlayer in the UK), the new (or innovative) offering is not only firmly tested against producing added value for the public, but also the damage existing commercial providers may suffer from seeing their number of audiences and therefore possibly advertising income reduced. To legitimate the introduction of these new services (or innovations of traditional ones), the BBC Trust—the governing body of the BBC between 2007 and 2017—had consulted license fee payers to ensure its decisions are properly informed by those who pay for the BBC.

While this process has proven a valid means to safeguard public accountability, three critical issues are unresolved (Meynhardt & Frantz, 2019; Murschetz, 2021a, b):

Firstly, as has long been argued by the commercial media sector, PSM occupies a privileged funding position that gives public media an unfair competitive advantage. To ensure that PSM are both accountable and transparent, many governments now insist on substantial reporting and monitoring of their activities, especially their financial dealings. Yet, such accountability is rarely shared and promoted to the public in a way that could build and maintain trust and credibility (PMA, 2019).

Secondly, while license fees remain by far the main source of funding for PSM in many countries across Europe and beyond (EBU, 2020b), discussion about the efficacy and governance of the model are back on the policy front burner (Puppis & Van den Bulck, 2018; Schweizer & Puppis, 2018).

Doubts on fee legitimacy and efficacy can be summarized as follows. Scholars (Puppis & Van den Bulck, 2018; Schweizer, 2020) suggest to:

- Transform the fee into a household charge (such as in Germany and Switzerland), replacing it with a specific fund ringfenced outside the state budget and financed by a PSM special tax (such as in Finland and Sweden), or simply replacing it with transfers from the state budget (as in North Macedonia, Norway and Romania) (EBU, 2020b)
- Change the methods for organizing fee collection, whereby the most widespread collection agencies are electricity suppliers and PSM themselves. Other possibilities witnessed are tax authorities, postal agencies, or private companies
- Directly correlating the fee volume with performance metrics (other than share and reach, but more quality-based PV indicators; see above).

Interestingly, the paradigm that PSM need to be financed through license fees still dominates academic discourses. Such a model, it is believed, would be the most appropriate, legitimate, and effective. However, the dominance of the license fee funding model needs to be more fundamentally reviewed, if not broken. The deficits of the model seem to outweigh its benefits (Fukumoto & Bozeman, 2019; Gransow, 2020; Kelly et al., 2002; Serong, 2017).

Given the situation in Austria, there is reason to claim that the fee model is outdated. While most European countries with a fee have moved away from the traditional criteria of radio/TV reception-devices ownership, Austria still firmly sticks to it. More so, in Austria the fee is also covering devices including a tuner, which remains a rather restricted definition of a reception device.

Thirdly, it appears that the level of the ORF's license fee is too high compared to the overall volume of public value content offered. In Austria, it needs a much stricter definition of what is public value media programming and what is not.

Principally, Para 31c of the ORF Act states that the funds received by ORF from the license fee may not be used in a manner which distorts competition which is not necessary for the fulfillment of the public service remit.

This includes the acquisition of broadcasting rights at excessive prices that are not justified according to commercial principles or the awarding of commercial communications at low prices, which only serve to increase the market share on the advertising market at the expense of competitors.

And fourthly, PSM must be able to work with stable multi-year budgets. However, the resulting social costs must be wisely balanced against the benefits of PSM offerings in the public interest. The PSM's budget, in turn, needed to be adjusted according to key measures alongside an evidence-based public value scorecard model. Multi-year budgeting is important and hence needs to be controlled by an independent regulator.

# 6 Conclusion

To conclude, this study has critically discussed public value theory and its applicability to Public Service Media (PSM) in the digital era. It asked whether PSM validly and evidently generate both economic value and societal welfare.

Given above results, three conclusions can be drawn:

Firstly, a clear definition of PSM's contribution to the public interest in the digital age is to refer to the central principles of public television, the public remit for full-service provision. Public service television is legally obliged to achieve geographic, program, and audience universality whereby it has something at some time to offer to everyone for the viewing nation as a whole. However, this means that public broadcasters are forced into an almost impossible balancing act: to survive they need to produce both distinctive niche products which the rest of the marketplace does not provide (including minority programming) and cater for a wide range of tastes and interests to justify the value of public funding. Therefore, viewer-financed public television must offer people something which they cannot otherwise see, serve minorities as well as fulfill the function of a national culture-identity carrier for majorities.

As a corollary, we doubt that the traditional concept of public value as widely scholarly agreed in PSM research is likely to effectively master the fundamental challenges of PSM. While accountability, diversity, excellence, independence, innovation, and universality are high-level objectives of PSM which are all legitimate, they are still rather elusive in character. They have high face validity, but only narrowly based explication, little consensus on meaning, and only recently emerging empirical verification. As for the future, we argue that to be truly effective, public value for PSM needs to be defined more broadly and narrowly at the same time. More broadly in mission and vision, and more narrowly in proven impact for society.

Secondly, PSM public value performance measurement is tricky and hard to come by. Therefore, benchmarking tools, vision and mission statement evaluations, input/output analyses, balanced scorecard techniques, cost–benefit analyses, and corporate social responsibility techniques need to be applied. Evidently, traditional reach and efficiency indicators alone will never grasp the scope and richness of PSM value to society. As argued and discussed above, a set of well-chosen criteria is needed to measure PSM performance adequately. Public value scorecard models, such as Meynhardt's presented here, are promising toolboxes for evaluating public value impacts of PSM.

Thirdly, we believe that PSM performance must be measured against well-chosen criteria. For the future, public value scorecard models, such as Meynhardt's PVSC presented above, are perfect toolboxes for evaluating public value impacts of PSM in times of the digital transformation (Meynhardt et al., 2017). Reviewing other frameworks—such as the BBC's RQIV (Reach, Quality, Impact, and Value for money) framework—elicit further valuable insights when evaluating the performance of PSM monitored against defined "drivers" of public value. Ultimately, it remains to be seen whether the ORF will succeed in transforming the country's media flagship to a real public value media platform. Evidently, it is not the only guarantor for the production and dissemination of public interest media communication in Austria. Other media also shape and create the public sphere in the Alpine republic, above all quality newspapers and non-commercial citizen media. Leading the ORF through the demanding and challenging transformation to remain a well-accepted and solidly valued PSM will depend on which public value criteria are chosen and balanced against value-driven performance indicators (Karmasin & Bichler, 2017).

## References


Alford, J., & O'Flynn, J. (2009). Making sense of public value: Concepts, critiques and emergent meanings. *International Journal of Public Administration, 32*(3–4), 171–191.

BBC. (2004). *Building public value. Renewing the BBC for a digital world.* BBC.

Benington, J. (2009). Creating the public in order to create public value? *International Journal of Public Administration, 32*(3), 232–249.

Benington, J. (2011). From private choice to public value? In J. Benington & M. H. Moore (Eds.), *Public value: Theory and practice* (pp. 31–51). Palgrave Macmillan.

Benington, J., & Moore, M. (2009). *Public value: The theory and practice.* Palgrave.

Bozeman, B. (2007). Public values and public interest: Counterbalancing economic individualism, .

Bozeman, B. (2009). Public values theory: Three big questions. *International Journal of Public Policy, 4*(5), 369–375.

Bozeman, B., Rimes, H., & Youtie, J. (2015). The evolving state-of-the-art in technology transfer research: Revisiting the contingent effectiveness model. *Research Policy, 44*(1), 34–49.

Bracci, E., Gagliardo, E. D., & Bigoni, M. (2014). Performance management systems and public value: A case study. In J. Guthrie, G. Marcon, S. Russo, & F. Farneti (Eds.), *Public value management, measurement and reporting* (pp. 129–157). Bingley.

Brookes, S., & Wiggan, J. (2009). Reflecting the public value of sport. *Public Management Review, 11*(4), 401–420.

Bryson, J. M., Crosby, B. C., & Bloomberg, L. (2015). Discerning and assessing public value: Major issues and new directions. In J. M. Bryson, B. C. Crosby, & L. Bloomberg (Eds.), *Public value and public administration* (pp. 1–21).

Cobbold, I., Lawrie, G., & Issa, K. (2004). Designing a strategic management system using the third-generation balanced scorecard. *International Journal of Productivity and Performance Management, 53*(7), 624–633.

Collins, R. (2007). The BBC and 'Public Value'. *Medien & Kommunikationswissenschaft, 55*(2), 164–184.

Conolly, L. (2013). Measuring public value: Reflections from Australia. In C. A. Scott (Ed.), *Museums and public value: Creating sustainable futures* (pp. 113–127). Routledge.

Deltenre, I. (2012). Do we need public service media? www.ebu.ch/news/2014/07/news-the-role-and-remit-of-public-ser.

Dobusch, L. (2018). Democratic media public spaces in the age of digital platforms. In K. Mitschka & K. Unterberger (Eds.), *Public open space. On the future of public media* (pp. 308–313). Facultas.

Donders, K., & Moe, H. (2011). *Exporting the public value test. The regulation of public broadcasters' new media services across Europe*. Nordicom.

Eberwein, T., Saurwein, F., & Karmasin, M. (2019). Public service broadcasting in Europe: A ratio-based comparison on the relationship between funding and audience performance. In Yearbook of Christian Social Sciences (Ed.), *Public service media* (pp. 141–167). Münster.

European Broadcasting Union EBU. (2020a). *Empowering society. A declaration on the core values of public service media.* : European Broadcasting Union.

European Broadcasting Union EBU (2020b). Licence fee 2020. www.ebu.ch/files/live/sites/ebu/files/Publications/MIS/members_only/funding/EBU-MIS_Licence_Fee_2020_public.pdf.

EBU – European Broadcasting Union. (2017). *Vision 2020*. European Broadcasting Union.

European Broadcasting Union. (2012). *Empowering society. A declaration on the core values of public service media*. European Broadcasting Union.

Faulkner, N., & Kaufman, S. (2018). Avoiding theoretical stagnation: A systematic review and framework for measuring public value. *Australian Journal of Public Administration, 77*(1), 69–86.

Fukumoto, E., & Bozeman, B. (2019). Public values theory: What is missing? *The American Review of Public Administration, 49*(6), 635–648.

Gransow, C. (2020). A question of value or further restriction? Public value as a core concept. In P. Savage, M. Medina, & G. F. Lowe (Eds.), *Universalism in public service media* (pp. 113–131). Nordicom.

Gundlach, H. (2011). *Public value in der Digital- und Internetökonomie*. Halem.

Heeks, R. (2008). Benchmarking e-government: Improving the national and international measurement, evaluation and comparison of e-government. In Z. Irani & P. Love (Eds.), *Evaluating information systems: Public and private sector* (pp. 257–301). Elsevier.

Hills, D., & Sullivan, F. (2006). *Measuring public value 2: Practical approaches*. Work Foundation.

Horner, L., Lekhi, R., & Blaug, R. (2006). *Deliberative democracy and the role of public managers*. The Work Foundation.

Karmasin, M., Süssenbacher, D., & Gonser, N. (2011). *Public value. Theorie und praxis im internationalen Vergleich*. VS Verlag.

Karmasin, M., & Bichler, K. (2017). Corporate social responsibility in the media industry: Setting the benchmark or falling behind? In K.-D. Altmeppen, C. A. Hollifield, & J. van Loon (Eds.), *Value-oriented media management: Decision making between profit and responsibility* (pp. 135–147). Springer.

Kelly, G., Mulgan, G., & Muers, S. (2002). Creating public value. An analytical framework for public service reform. .

Knoll, E. M. (2012). *The public value notion in UK public service broadcasting: An analysis of the ideological justification of public service broadcasting in the context of evolving media policy paradigms*. London School of Economics and Political Science (LSE). PhD thesis.

Mazzucato, M., & Semieniuk, G. (2017). Public financing of innovation: New questions. *Oxford Review of Economic Policy, 33*(1), 24–48.

Mazzucato, M., Conway, R., Mazzoli, E., Knoll, E., & Albala, S. (2020). *Creating and measuring dynamic public value at the BBC*. UCL Institute for Innovation and Public Purpose, Policy Report (IIPP WP 2020-23).

Mazzucato, M. (2016). From market fixing to market-creating: A new framework for innovation policy. *Industry and Innovation, 23*(2), 140–156. https://www.tandfonline.com/doi/full/10.1080/13662716.2016.1146124

McQuail, D. (2010). *McQuail's mass communication theory* (6th ed.). Sage.

Meynhardt, T. (2009). Public value inside: What is public value creation? *International Journal of Public Administration, 32*(3–4), 192–219.

Meynhardt, T. (2015). Public value—turning a conceptual framework into a scorecard. In J. M. Bryson, B. C. Crosby, & L. Bloomberg (Eds.), *Public value and public administration* (pp. 147–169). Georgetown University Press.

Meynhardt, T. (2019). Public value: Value creation in the eyes of society. In A. Lindgreen, N. Koenig-Lewis, M. Kitchener, J. D. Brewer, M. H. Moore, & T. Meynhardt (Eds.), *Public value: Deepening, enriching, and broadening the theory and practice*. Routledge.

Meynhardt, T., & Bartholomes, S. (2011). (De)composing public value: In search of basic dimensions and common ground. *International Public Management Journal, 14*(3), 284–308.

Meynhardt, T., & Frantz, E. (2019). Media public value as a contribution to the common good. *Media Perspectives*, 444–451.

Meynhardt, T., & Jasinenko, A. (2020). Measuring public value: Scale development and construct validation. *International Public Management Journal, 24*(2), 222–249.

Meynhardt, T., Brieger, S., Strathoff, P., Anderer, S., Bäro, A., Hermann, C., Kollat, J., Neumann, P., Bartholomes, S., & Gomez, P. (2017). Public value performance: What does it mean to create value in the public sector? In R. Andeßner, D. Greiling, & R. Vogel (Eds.), *Public sector management in a globalized world* (pp. 135–160). Springer Gabler. https://doi.org/10.1007/978-3-658-16112-5_8

Moore, M. (1995). *Creating public value: Strategic management in government*. Harvard University Press.

Moore, M. H. (2013). *Recognizing public value*. Harvard University Press.

Mulgan, G., Breckon, J., Tarrega, M., Bakhshi, H., Davies, J., Khan, H., & Finnis, A. (2019). *Public value. How can it be measured, managed and grown?* NESTA.

Murschetz, P. C. (2021a). Geschäfts- und Erlösmodelle in den Medien. In J. Krone & T. Pellegrini (Eds.), *Handbuch Medienökonomie. Springer reference Sozialwissenschaften*. Springer VS. https://doi.org/10.1007/978-3-658-09632-8_19-3

Murschetz, P. C. (2021b). The future of ORF in the digital age. Three future scenarios and the question of fee financing of public service broadcasting in Europe. In P. C. Murschetz, N. Alm, F. Weder, & M. Friedrichsen (Eds.), *The digital transformation of the media*. Springer-Gabler. in print.

ORF. (2017a). Independence. Quality. Diversity. 12 theses on Austria as a media location. Independence. Quality. Diversity: ORF Director-General Wrabetz presents 12 Theses on Austria as a Media Location—der.ORF.at

ORF. (2017b). Public Value Bericht 2017/18—"Das Handbuch". ORF Public Value.

ORF. (2019). Public Value Bericht 2018/2019—"Der Auftrag 2019—für Europa"! pvb_201819_i_austria.pdf (orf.at)

ORF. (2021). ORF TransFORM—Der Bericht. ORF Public Value—ORFTransFORM-PublicValue Report 2020/21.

PMA—Public Media Alliance. (2019). *Funding public media an insight into contemporary funding models*, by Mervyn Warner, www.publicmediaalliance.org/insight-funding-public-media/.

Puppis, M., & Van den Bulck, H. (2018). Finanzierungsmodelle und Unabhängigkeit des öffentlichen Rundfunks. In K. Mitschka & K. Unterberger (Eds.), *Public open space. Zur Zukunft öffentlich-rechtlicher Medien* (pp. 257–271). Facultas.

Schweizer, C. (2020). *Öffentliche Rundfunkorganisationen als Media Commons? Vorschlag eines neuen Leitbegriffs und vergleichende Analyse in 16 Ländern*. Nomos.

Schweizer, C., & Puppis, M. (2018). Public service media in the 'Network' era. A comparison of remits, funding, and debate in 17 countries. In G. F. Lowe, H. Van den Bulck, & K. Donders (Eds.), *Public service media in the networked society. RIPE@2017* (pp. 109–124). Göteborg. http://www.nordicom.gu.se/sites/default/files/kapitel-pdf/07_schweizer_puppis.pdf

Serong, J. (2017). The public broadcasters and public value. On the untapped potential of the concept of public value. *Communicatio Socialis, 52*(2), 20–35.

Spano, A. (2014). How do we measure public value? From theory to practice. In J. Guthrie, G. Marcon, S. Russo, & F. Farneti (Eds.), *Public value management, measurement and reporting* (pp. 353–373). Emerald.

Talbot, C. (2009). Public value—the next "big thing" in public management? *International Journal of Public Administration, 32*(3–4), 167–170.

Talbot, C., & Wiggan, J. (2010). The public value of the National Audit Office. *International Journal of Public Sector Management, 23*(1), 54–70.

Van den Bulck, H. (2015). Public service media accountability in recent decades: A progressive shift from state to market. In K. Arriaza Ibarra, E. Nowak, & R. Kuhn (Eds.), *Public service media in Europe. A comparative approach* (pp. 73–89). Routledge.

Van Dijck, J., Poell, T., & de Waal, M. (2018). *The platform society. Public values in a connective world*. Oxford University Press.

WEF—World Economic Forum. (2020). *Understanding value in media: Perspectives from consumers and industry*. World Economic Forum Publication.

# Crisis Communication and Corporate Responsibility in Media Companies

## A Case Study of the “Relotius Scandal” and an Exploratory Study of Communication Managers and Ombudsmen

**Christian Pieter Hoffmann and Stephan Russ-Mohl**

## 1 Introduction: The Nexus of CSR and Crisis Communication

Media companies play a key role in bolstering corporate social responsibility (CSR): Media coverage has been shown to positively relate to CSR strength (Zyglidopoulos et al., 2012) and to motivate corporate leadership to focus on CSR (Garcia-Sanchez et al., 2014). Companies with higher levels of CSR receive more favorable media coverage (Cahan et al., 2015). Through high-quality reporting, media companies contribute to corporate responsibility. Put differently, the corporate responsibility of media companies extends to the quality of their journalism. In journalism and political communication studies, this argument has been extended beyond a business context. High-quality journalism is frequently related to the quality of democratic institutions (Strömbäck, 2005). Hence, journalistic quality is a critical element of media companies' corporate social responsibility.

A context, in which the nexus between the corporate responsibility of media corporations, CSR communication, and journalism is of particular interest is crisis communication. Coombs (2007, pp. 2–3) defines a crisis as “the perception of an unpredictable event that threatens important expectancies of stakeholders and can seriously impact an organization's performance and generate negative outcomes.” Importantly, this definition highlights that crises may exert a deleterious influence on

C. P. Hoffmann (✉)
Institute of Communication and Media Studies, University of Leipzig, Leipzig, Germany
e-mail: christian.hoffmann@uni-leipzig.de

S. Russ-Mohl
Kleinmachnow, Germany
e-mail: stephan@russ-mohl.de

 
F. Weder et al. (eds.), *CSR Communication in the Media*, CSR, Sustainability, Ethics & Governance, https://doi.org/10.1007/978-3-031-18976-0_6

a corporation's performance, the level of quality in products or services provided to stakeholders. Applied to media corporations, this implies that a corporate crisis could negatively affect the quality of journalism provided—and thereby media company's corporate social responsibility. Crisis communication is therefore intertwined with CSR communication.

This chapter will examine crisis communication in media companies based on a case study and an interview study among communication managers of leading German media companies and ombudsmen. In particular, it will explore how corporate crises affect journalistic quality, the delineation of journalism and corporate communication in media companies, and the role of institutions ensuring journalistic quality (here: ombudsmen) in crisis management. The case study will focus on the "Relotius scandal," named after a prominent journalist at the leading German news magazine DER SPIEGEL who was convicted of serial fabrications (Hoffmann & Russ-Mohl, 2020). We will show that in media companies, crises can engender a blurring of journalism and corporate communication, which potentially affects journalistic quality—and thereby media companies' corporate responsibility. Conversely, we will discuss how institutions ensuring journalistic quality, such as ombudsmen, can be a boon to both journalism and corporate responsibility.

## 2 Crisis Communication in Media Companies

To date, few studies have focused on corporate communication or PR in media companies. Studies in the German speaking world have focused primarily on product and market communication, advertising, and branding (Gündling, 2007; Eble, 2013; Holtzhausen, 2015). Some case studies outline the PR efforts of specific media companies. One of the pioneering contributions in this research stream is Fröhlich's "Rundfunk-PR im Kontext" (Broadcast PR in Context) (1994). Fröhlich points out that journalists resent being integrated into media companies' PR or marketing efforts. The separation between newsrooms and media management is regarded as an important journalistic value. On the other hand, journalists frequently engage in discourse with audiences and thereby shape the public perception of media corporations.

The traditionally contentious relationship between journalism and PR may explain why little attention is paid to the corporate communication of journalistic organizations. As recently as 2000, Wehmeier and Bentele noted a low level of professionalization of PR in media companies. Arnold (2018), however, notes a surge in professionalization just a few years later, specifically in the context of digitalization. For example, a majority of German media companies now feature a press section on their websites where contact persons are named and press releases and other materials are made available (for example, in the so-called media libraries). Funck (2015) also notes that newspaper publishers are making increasing efforts to establish transparency—but her analysis focuses on editorial channels and not public relations in the narrower sense.

The authors could not identify a single study expressly examining crisis communication in media companies (more precisely: journalistic organizations, as Weiss-Blatt (2021) explores the crisis communication of social media companies). Conversely, however, in journalism research, numerous studies discuss journalistic quality and institutions to ensure the quality of journalistic products (cf., Meier, 2019). Some institutions bolstering journalistic quality reside outside of media corporations, such as education programs, professional associations, law and codices, market structures and competition, etc. At the same time, journalistic institutions implement structures and processes that aim at ensuring output quality, such as editorial proofreading, fact-checking, sheet reviews, or ombudsmen (also: reader representatives, public editors, or readers' advocates). Ombudsmen serve as boundary-spanners, representing readers' views and feedbacks within journalistic organizations (Ferrucci, 2019). Frequently, they also serve to increase the transparency of journalistic work toward stakeholders.

To summarize, while media companies engage in corporate communication to create value and ensure legitimacy (Zerfaß, 2014; Bachmann & Ingenhoff, 2017), little research has been conducted on the particularities of these organizations. To date, no study has addressed crisis communication in media companies. Some studies indicate that corporate communication in media companies has increased in professionality—up to and including sustainability and financial communication (Trommershausen, 2011; Laskin & Samoilenko, 2014). A lack of communicative professionalism invariably represents a potential source of crisis—especially in the age of digital media. However, in the context of media companies, corporate and crisis communication—just as CSR—cannot be detached from journalistic quality. Based on a case study of the "Relotius scandal," we will examine how a crisis caused by lack of journalistic quality is addressed in a journalistic organization, the role of journalism in crisis communication, and the potential contribution of ombudsmen as an institution of quality control to crisis management.

## *2.1 Methodology*

The present study comprises three analytical steps. First, reporting on the "Relotius scandal" was reviewed to analyze the course of the crisis and, in particular, the communication of the media company concerned. Next, interviews were conducted with representatives of the SPIEGEL editorial leadership and management as well as communication managers of major German media companies. Since one of SPIEGEL's responses to the crisis was to install an ombudsman's office, a third step, based on an explorative qualitative survey of ombudsmen, was conducted to examine their position and role in crisis management.

The media analysis focused primarily on industry and trade media that were surveyed for mentions of the Relotius crisis in the period 2019/2020. In addition, SPIEGEL's communication on the matter was analyzed, including the independent investigation commission's final report. Interviews were conducted with SPIEGEL's editor-in-chief Steffen Klusmann and the group's head of corporate communication, Anja zum Hingst. At the end of 2020 and beginning of 2021, supplementary interviews were conducted with active and former communication managers of major German media companies.[1] While the interview participants agreed to be named, the analysis and presentation of results were carried out anonymously. The ensuing ombudsmen survey addressed, in addition to the SPIEGEL ombudsman's office, the contacts registered in the Association of Media Ombudsmen (VDMO) (22 personal members and five institutional members). Five ombudsmen from major German regional newspapers agreed to participate. Survey participants were assured of anonymity. The results of the three analytical steps are presented successively below.

# 3 SPIEGEL's Crisis Case

## *3.1 The Relotius Scandal*

The most detailed review of the "Relotius scandal" was published by the affected organization itself: the "Final Report of the Independent Investigation Commission" (Fehrle et al., 2019). This fact is indicative of the crisis communication of both the SPIEGEL Group and other media organizations in the course of the scandal. In 2018, the SPIEGEL editor-in-chief and management simultaneously publicized the scandal while announcing that it would set up an independent investigation commission (cf. Author Not Named, 2018a). As a result, the affected organization was able to investigate the "Relotius case" more comprehensively than any other editorial office due to its privileged access to data.

Claas-Hendrik Relotius had been considered a particularly promising journalistic talent. The final report of the SPIEGEL investigative commission speaks of around 40 awards he had received for his reporting (cf. Fehrle et al., 2019). The downfall of the prominent and widely admired reporter came about when he started cooperating with Juan Moreno, a "permanent freelancer" on the Berlin editorial staff of SPIEGEL who, by his own admission, hardly knew Relotius before they began working together. They cooperated on a report about a US vigilante group on the Mexican border ("Jaeger's Grenze") during which Moreno recognized discrepancies in the information provided by his colleague and realized that many of his claims in the manuscript seemed implausible.

[1]Participants in the interview study were: Steffen Grimberg (former ARD spokesman at Mitteldeutscher Rundfunk (MDR) as part of the ARD chairmanship), Oliver Herrgesell (EVP Communications WarnerMedia Sales & International), Christian Körner (Head of Communications and Corporate Spokesperson, Mediengruppe RTL Deutschland), Silvie Rundel (Head of Corporate Communications and Events ZEIT Publishing Group), Ingrid Schmitz (Corporate Spokesperson and Head of Communications, WDR), Frank Thomsen, (Head of Corporate Communications, Gruner + Jahr).

Moreno brought these discrepancies to the attention of the head of the Society desk in mid-November 2018. According to his own recount, this met with a restrained response. In a telephone conversation, Moreno was threatened with dismissal if his accusations proved to be false (cf. Author Not Named, 2018b). Moreno eventually traveled to the locations of the report on his own and discovered that Relotius had quoted protagonists in his text without having spoken to them. These internal findings moved the SPIEGEL leadership to publicly acknowledge the scandal—and to preempt any reporting by competing outlets.

SPIEGEL publicized the forgery scandal while its editorial team initiated a review of all texts contributed by Claas Relotius (Fichtner, 2018). A total of 59 articles were examined for forgeries—many containing mistakes or misleading statements, some downright fabrications. In most cases, these fabrications served to present a riveting and politically consonant narrative. The scandal also affected a number of other media organizations that had cooperated with Relotius in the past. The industry magazine *Journalist* reported in March 2019 that out of 13 editorial offices or editorial networks that had published texts by Relotius, eight had discovered forgeries (Lungmus, 2019).

On May 24, 2019, SPIEGEL's management presented its investigation commission's final report. The managing director and editor-in-chief wrote: "The good news: No evidence was found that anyone in the company knew about the forgeries, covered them up, or was even involved in them. The bad news: We let ourselves get sucked in by Relotius and made mistakes to an extent that, measured against the standards of this house, are unworthy" (Hass & Klusmann, 2019, p. 130).

The final report also analyzed causes of the forgery scandal. It does not indicate that Relotius acted with particular sophistication or criminal ingenuity, but rather with social intelligence. Relotius was a popular and almost revered colleague, especially in his own department, who openly approached co-authors, fact-checkers, and even readers. Essentially, the commission's investigation revealed a systemic failure. The leadership of the Society desk valued the stylistic form of free-wheeling reportage and a successful narrative over the truth and was inclined toward exaggeration and dismissive of criticism. This made it easy for Relotius to "get away with it" (Fehrle et al., 2019, p. 139).

As a result of the scandal, an undisclosed number of lower-level journalists and staff left SPIEGEL (Hass & Klusmann, 2019, p. 130). The editorial leadership was reshuffled, the responsible head of the Society desk was deprived of an imminent promotion, yet not dismissed. After some delay, the Society desk was restructured following a proposal by the independent commission. Finally, SPIEGEL announced the institutionalization of an ombudsman.

## 3.2 SPIEGEL's Crisis Communication

A key peculiarity of the case described is that the media company involved was largely able to retain control over the coverage of the scandal. The offense was discovered by and within the organization. The decision to go public was made by the editorial board, not the corporate communications department, and there was never a press release from the SPIEGEL Group disclosing the "Relotius case." The magazine, however, published various pieces of editorial content. With the exception of write-ups on personnel changes, all external crisis communication was conducted via channels of the editorial office.

The examination of the scandal was not delegated to external, independent actors; it remained essentially in the hands of the editorial staff concerned, indeed in some cases in the hands of editors who were directly involved in the scandal. In the process, there were slight embellishments and manifest attempts at framing, which are common in the context of crisis communication. Usually, however, they are also quickly uncovered and corrected by journalists researching the respective case. In the "Relotius case," instead, the most relevant coverage was provided by the news organization affected by the scandal.

Analyzing SPIEGEL's crisis communication in this case, some features are of particular notice: (1) SPIEGEL assumed some responsibility by pointing out the failure of its own editorial processes. However, the responsible protagonists named in the final report of the investigation commission did not face any noticeable sanctions. (2) Relotius is consistently presented as the main culprit, he is portrayed as a resourceful fraudster who outwitted a robust system. As a result, the editorial staff declared itself a victim of the perpetrator—sometimes in an emotive manner: "We love SPIEGEL, and we are sorry that we could not spare the publication—our dear friend—this crisis" (Klusmann & Kurbjuweit, 2018). (3) Throughout SPIEGEL's communication on the crisis it was repeatedly pointed out that the magazine was not the only "victim" of Relotius, but that there were numerous other editorial offices as well as prize committees and juries who had fallen for the deceit. This could be read as a warning to other outlets to be too quick to point fingers.

What is most remarkable and unusual about SPIEGEL's crisis communication is the means employed: As described, communication was conducted almost exclusively via the editorial staff. Thus, in the case of the Relotius crisis, an asymmetry usually inherent in crisis communication was eliminated: The affected organization struggles to communicate as objectively as possible and strives for transparency—journalists reporting on the case classify, evaluate, and repeatedly also exaggerate (cf. Bentele, 1994). The pronounced framing of the crisis on the part of SPIEGEL anticipates and internalizes the latter task of journalism. The language and stylistic devices used in SPIEGEL's communication on the crisis were ornate, emotive, and dramatic in a way that would usually be considered improper in other cases of crisis communication—not least because such language would be criticized by the contributing journalists (cf. Töpfer, 2006).

From SPIEGEL's perspective, employing editorial channels for crisis communication proved advantageous. The stylistic and framing opportunities as well as the editorial space available made it possible to convey the perspective of the affected organization in a unique way. SPIEGEL stylized itself as the victim of an ingenious fraudster. This victim status conveyed in emotional language and via editorial content can also be interpreted as an appeal to stakeholders to build up sympathy and a sense of shared admiration for SPIEGEL as an institution. SPIEGEL was also able to diffuse responsibility by pointing out how other editorial offices and award juries were affected and shared liability. Various major editorial offices in Germany had to put their own house in order—which may have caused them to abstain from further investigations into SPIEGEL's misconduct.

## 4 Crisis Communication in Media Companies from the Perspective of Communication Managers

As noted, representatives of SPIEGEL were interviewed on their perceptions and evaluations of the publisher's crisis communication in the "Relotius case." Anja zum Hingst, Head of Communications and Marketing at SPIEGEL Group, reports that the publisher's communications team, consisting of three members in addition to herself, had not been able to rely on a pre-existing emergency or crisis response plan. A strategy of "total transparency" was decided on the spot by the editor-in-chief in conjunction with management. A crisis team was convened, which decided by consensus that external communication would essentially take place through the channels of the editorial office. Editor-in-chief Steffen Klusmann justifies this by saying that Relotius had also been part of the editorial team and that the scandal primarily concerned editorial contributions. Zum Hingst emphasizes the advantage of focusing all crisis communication to all stakeholders on one channel and emanating from one source.

Inquiries from other editorial departments, however, were forwarded to the communications department. As a rule, they were answered personally by Ms. zum Hingst—always in coordination with the crisis team. Ms. zum Hingst states that SPIEGEL Group's crisis communications succeeded in staying "ahead of the game" and largely setting the pace—among other things, by setting up the independent investigation commission, by transparently disseminating its findings and also, beyond that, by uncovering and publicizing the case.

Interviews with communication managers of other leading German media corporations revealed that the spokesperson role in such organizations is often occupied by a former journalist. The interviewees emphasize that personal familiarity with the journalistic way of working and thinking is helpful for their job in several ways: First, there are often personal relationships with former colleagues, which can be helpful when it comes to initiating publications or correcting mistakes and misunderstandings "through official channels." Second, personal experience makes it easier to empathize with the needs of the target group. Third, and importantly, familiarity with the journalistic perspective also facilitates collaboration with in-house editorial staff.

Interviewees, however, also emphasize that journalism's widespread skepticism toward public relations can create antagonism toward in-house corporate communications. "Media companies have an especially high number of media experts.

Anyone who can read a newspaper also thinks they can do public relations. In other companies, there may be more respect for the public relations role than in media companies. Not because public relations is perceived as less important, but because there are so many people who think they know how to do it anyway" (IP2).

Several interviewees report that for a long time, the respective institutions pursued a rather passive and reactive communications strategy and responded primarily to inquiries. This has changed considerably in the past decade for various reasons. On the one hand, media houses are publicly criticized more often and are under pressure to legitimize their business. On the other hand, digital media open up new opportunities for direct targeting, corporate publishing, and content marketing. "We generate much more content ourselves and are no longer that dependent on having to talk to journalists" (IP3). This confirms that digitalization contributes to the professionalization of corporate communication in media companies.

The interviewees find that various editorial departments have cut back on media departments and the corresponding specialist journalists in recent years. However, a number of relevant trade media still exist. They speculate that journalists naturally have a high level of interest in their own industry. "In media companies, the attention for media companies is different, there are more mutual dependencies" (IP2). The economic disruption of the media industry in the wake of digitization is seen as a recurring cause for crisis communication. Mergers, sales, rounds of cost-cutting, and layoffs are often the subject of critical reporting. "This personal suffering is reflected in how stories are reported on, which can make our job difficult sometimes" (IP3).

Similar to the "Relotius scandal," a recurring flashpoint for media companies is the behavior of in-house editorial staff. Attempts to deceive are an exception here; more often, journalists' clumsy or provocative statements on social media (especially Twitter) have to be intercepted communicatively. Public broadcasting in particular is under public criticism from various interest groups due to its programming and journalistic output.

A critical component of the communication management in media companies is the interaction with in-house editorial departments. This exchange is described as not only simultaneously close and trusting, but also professionally distanced. "Editorial departments tend to report more critically on their own house in order to prove their own independence" (IP2). "Nevertheless, colleagues naturally approach their own company differently—I don't want to say with kid gloves, but with greater caution than other editorial departments outside do" (IP1). Traditionally, great importance is attached to the independence of the editorial office from company management. At the same time, interview partners also note that economic constraints may weaken this independence, for example, when the company's own editorial channels are increasingly used for marketing purposes.

In common practice, in-house editorial departments are served on an equal footing and usually at the same time as external editorial departments. If the company wants to engage in agenda setting, those responsible for communications even prefer to work with external editorial offices in order to give the reporting greater credibility. However, these norms change in the event of a crisis, especially if an in-house editorial department is affected, such as in the "Relotius scandal." In this case, internal "colleagues own the high ground" (IP5).

In general, crisis preparedness is weak among the media companies surveyed. Media training sessions with board members are most frequently cited as a familiar tool for crisis preparation. Crisis training sessions rarely take place and crisis response processes are often spontaneous and rather informal: "Actually, everything happens via WhatsApp. Someone shares the message along the lines of: something has happened here. Then we form a group that includes all the relevant decision-makers, if it doesn't already exist anyway. And then, together with the management, the editorial management, us, everyone involved, we decide how to react. We also address the question of whether we report on it ourselves" (IP5).

Interviewees universally praise SPIEGEL's crisis communication in the Relotius case. The strengths of the approach described above are repeatedly emphasized with approval. These include the investigative work carried out internally and by the independent commission which enabled the company to largely set the agenda itself: "They produced a simplified presentation, they were self-determined and they were transparent. (...) SPIEGEL's narrative is *the* valid version" (IP2). "I think it's exemplary. Corporate communications was certainly heavily involved. But they chose a good form [editorial channels] there. It's just also much more impressive than a press release" (IP6).

The communications experts interviewed are thus aware that media companies have an advantage in the event of a crisis: in-house editorial channels with a wide reach and (relatively) high credibility. This gives media companies advantages in control over messaging and framing, not least because other newsrooms are willing to follow the reporting of the newsroom concerned if there is credible transparency. Critically, it should be noted that the interviewed experts largely focus on the ensuing media response. Less attention is paid to the effects of a crisis resulting from editorial (mis)behavior on the audience. Is there a loss of audience in the event of a crisis? What are the implications for the reach and credibility of the editorial channels employed? Does trust in journalistic media erode? To counteract such consequences, SPIEGEL announced that in reaction to the Relotius crisis it would set up an ombudsman's office.

## 5 Ombudsman Offices as Instances of Communication and Crisis Management

### *5.1 The SPIEGEL Ombudsman's Office*

The SPIEGEL ombudsman office currently consists of two editors—a documentary journalist and the head of news—as well as the head of the legal department at SPIEGEL. It is thus better equipped than most comparable offices—usually ombudsmen are "lone wolves." According to the ombudsman's office, it does not keep statistics on the complaints it receives. One of the reasons for this is that it is difficult to classify individual cases: from serious indications of genuine misconduct to submissions that are more likely to be dealt with by the legal department, from mere but sincere criticism, to insults of the editorial team or irrelevant nonsense.

The ombudsman's office was set up jointly by the editor-in-chief and management. The fact that "no delegation to individual members takes place in the case of serious submissions, but rather a six-eye principle applies" ensures "the greatest possible independence, especially since two of the three members are not hierarchically subordinate to the editor-in-chief." The ombudsman's office is "organizationally a mix of journalistic quality assurance and compliance architecture." As such, the ombudsmen office straddles the line between editorial office and management, it reports to management and the editors-in-chief. If the journalistic violation becomes a threat to the SPIEGEL brand, management is responsible.

SPIEGEL is not the first media company to react to a scandal by setting up an ombudsman's office. The most prominent case to date was the New York Times, which appointed its Public Editor in 2003 following the Jason Blair forgery scandal—a similar body that was able to operate until 2017.

According to different accounts (Russ-Mohl, 1994; Elia & Russ-Mohl, 2007, Okrent, 2008), the following "ideal-type" requirements are directed at a media ombudsman:

- They must be experienced in journalism and should know the editorial department well, which supports their role as a critical authority and as a mediator.
- They should be internally and economically independent—especially from the editor-in-chief and the managing editors, whose trust they should nevertheless enjoy. Management of the media company should decide on their employment and contract—which of course limits their independence.
- Beyond their own journalistic experience and authority, they must bring media competence to bear in order to be able to deal with complaints fairly and openly, to discuss them with the editors concerned and to settle disputes with a chance of success.
- They should be visible both internally and externally, both as a contact person and as a mediator who publicizes their opinions ideally with their own column.
- In order to gain trust, their explanatory function is particularly important for third parties: conflict cases provide an ideal background to make journalistic operations transparent.

Measured against this standard, the setup of the SPIEGEL ombudsman's office fulfills the first three criteria: The three-member body incorporates journalistic experience supplemented by legal experience, it knows the editorial staff well and presumably enjoys their trust. It also has a high degree of journalistic independence, at least if the three members are in agreement. However, the opportunity to make the work of the ombudsman's office visible to the "outside" was deliberately foregone.

## 5.2 *The Role of Ombudsmen in Crisis Prevention and Response*

For the most part, the newspaper ombudsmen interviewed for this analysis are more visible than their new colleagues from SPIEGEL, who are only referred to in the small print of the imprint. But there are also considerable differences among the regional newspapers: In one case, the ombudsman complains that there is "no reference to the complaints office" on the homepage; in another, it is mentioned daily. In a third and fourth example, there is a regular and clearly visible reference to the complaints office in the ombudsman column. On the other hand, there is also a regularly recurring comment column, in which "various complaints and reader concerns" are taken up "as well as the editorial work is explained," without explicitly referring to the ombudsman.

The number of complaints handled is not recorded by all ombudsmen. One ombudsman reports that he received "a great many reader inquiries and also complaints," but was only asked specifically for mediation assistance five times in 2020. A second and third interviewee stated that they had received an estimated 150 and 200 letters, respectively, over the course of the year. The fourth interviewee reported receiving around 20 to 30 complaints and questions a week—or between 1000 and 1500 over the course of a year. "Sometimes there are no complaints or inquiries for a week, sometimes 10 to 15."

Complainants are rarely deemed fully justified. One of the interviewees reported that he had done so only once in 2020; in all other cases, he had explained and defended the journalistic work of the editorial office. "Often an explanation of the editorial work, the processes and pressures in the editorial office is then sufficient for the readers to withdraw their complaint." One ombudsman highlights his work behind the scenes: "From time to time, I criticize content or make recommendations, even on a closed network for editorial staff members on Facebook. At times, discussions occur there. However, these are rare."

Transparency is regarded as the main advantage of an ombudsman's office for a media company: "The editorial team has begun to work more transparently. There has been a growing willingness on the part of the editorial staff to take reader complaints seriously, to deal with them and to respond as profoundly as possible. The likelihood of legal and/or ethical errors is dwindling," states one interviewee. In concrete terms, press council reprimands could also be prevented if errors were made clear in advance. Another interviewee believes that the greatest advantage is "of course to the reader," but that this is "good for the image of our company"—"-because the reader advocate is someone who has the time to deal with concerns without the pressure of day-to-day business."

From the perspective of crisis management, the institution of an ombudsman office can largely be characterized as a crisis prevention approach. Internally, it increases awareness for journalistic quality, and for audience feedback. Externally, it increases transparency and helps stakeholders understand journalistic practices, fostering trust. The latter advantage, of course, only occurs if the ombudsman receives sufficient visibility in the journalistic output. A well-respected ombudsman office, therefore, should decrease the likelihood of crises due to editorial misconduct. From the interviewees' perspective, media corporations without an ombudsman office "are foregoing a great opportunity to strengthen their credibility: through transparency, self-reflection and the ability to take criticism. They are forfeiting an additional opportunity to enhance journalistic quality."

## 6 Concluding Remarks

The "Relotius scandal" serves as a flashpoint to examine crisis communication practices in media corporations—and their relationship to CSR. While in many industries, CSR issues can be the cause of corporate crises, among media organizations, crises are much more likely to arise from editorial misconduct or simply journalists' (mis-)behavior. As noted above, such crises restrict organizations' room for maneuver and may affect product or service quality. Crises, therefore, constitute a challenge to corporate responsibility. In the case of media companies, journalistic quality is at the heart of social responsibility—both in relation to businesses, but also other powerful institutions and public discourse in general. Therefore, it is important to understand how crises and crisis communication relate to journalistic quality.

Our analysis highlights how in the case of the "Relotius scandal," the affected organization relegated all relevant crisis communication to editorial channels, largely delegating responsibility from the communications department to the editorial staff. As our interviews show, this approach is widely lauded and regarded as exemplary by other communications professionals in the industry. Indeed, our analysis reveals a number of advantages of the chosen approach: opportunities to stay "ahead of the fray" and set the agenda, reach and copious editorial space to put forth the organizations point of view, extensive framing of the crisis. We also observe notable restraint among other news outlets in their reporting in the scandal, which may be due to the particularities of this case. SPIEGEL's crisis communication is seen as transparent, credible, and comprehensive by the interviewed experts.

On the other hand, it is notable that the well-established and cautiously guarded separation of editorial staff and management blurs and almost disappears in this case of a crisis caused by editorial misconduct. The editorial staff largely takes over the role of spokesperson and represents the corporate position. Under normal circumstances, this would be regarded as a severe threat to editorial independence and journalistic quality. Editorial staff and management form a joint crisis response team and closely coordinate all external and internal communication. This supports the insight that crises restrict organizational leeway and pose a threat to product or service quality. To put it more bluntly: If it is indeed "best practice" among media corporations for editorial staff to serve as spokespersons in the case of a crisis originating from the organization's journalism, then such crises pose a serious challenge to media corporations' corporate social responsibility.

As we have seen, in the case of the "Relotius scandal," SPIEGEL's reporting was far from unbiased, it engaged in diffusion of responsibility, self-victimization, ingratiation—quite defensive crisis response tactics (Coombs, 2007). Both framing and priming are evidence in SPIEGEL's narration. All of these tactics may be common in crisis communication, they do, however, challenge norms of journalistic quality and may endanger the outlet's credibility. It is of particular interest, therefore, that the organization simultaneously decided to invest in an institution bolstering journalistic quality and transparency, an ombudsman office. In this instance, the implementation of such an institution can be criticized as half-hearted, as the office does not enjoy any outside visibility and is not systematically integrated into the organization's journalistic output (i.e., a column).

Despite this shortcoming, we identify the ombudsman office as a promising component of both crisis prevention and media companies' CSR efforts: It constitutes an institution that systematically increases the salience of journalistic quality and stakeholder expectations. It potentially increases transparency toward stakeholders and bolsters trust and credibility. It should be noted that we focus on the ombudsman in this instance as it was part of SPIEGEL's crisis response. Our insights, however, can be extended to other institutions ensuring journalistic quality within media organizations—especially if they serve a boundary-spanning function to stakeholders.

To summarize, this contribution highlights the complex interdependence of journalistic quality, corporate responsibility, and crisis management/communication in media companies. It is noteworthy that little research has been dedicated to the corporate communication of media companies, specifically crisis communication. We show that media companies enjoy some advantages, but also some specific challenges when it comes to crisis communication. Also, media companies can be considered increasingly prone to crises in the digital age (Russ-Mohl, 2017). Their corporate communications departments are usually not as well-resourced and crisis-experienced as comparable departments in other industries. More research is therefore required to deepen our understanding how crises challenge both journalistic quality and media companies' corporate social responsibility.

## References

Arnold, D. (2018). Das Vertrauensdilemma im Verhältnis zwischen Medien-PR und Medienjournalismus. In O. Hoffjann & R. Seidenglanz (Eds.), *Allmächtige PR, ohnmächtige PR* (pp. 175–196).

Author Not Named. (2018a). Der Fall Relotius. Kommission aus erfahrenen Journalisten soll Routinen beim SPIEGEL hinterfragen (2018, 19. Dezember). https://www.SPIEGEL.de/kultur/gesellschaft/der-fall-claas-relotius-kommission-aus-erfahrenen-journalisten-soll-routinen-beim-SPIEGEL-hinterfragen-a-1244680.html. Accessed 17 July 2019.

Author Not Named. (2018b). Betrugsfall Claas Relotius. "SPIEGEL"-Reporter Moreno wurde offenbar mit Rauswurf gedroht (2018, 22. Dezember). https://www.faz.net/aktuell/feuilleton/

medien/fall-relotius-SPIEGEL-reporter-mit-rauswurf-gedroht-15954983.html. Accessed 17 July 2019.
Bachmann, P., & Ingenhoff, D. (2017). How do media companies gain legitimacy? An experimental study on the (Ir)relevance of CSR communication. *International Journal of Strategic Communication, 11*(1), 79–94.
Bentele, G. (1994). Öffentliches Vertrauen—normative und soziale Grundlage für Public Relations. In W. Armbrecht & U. Zabel (Eds.), *Normative Aspekte der Public Relations. Grundlegende Fragen und Perspektiven* (pp. 131–158). Eine Einführung.
Cahan, S. F., Chen, C., Chen, L., & Nguyen, N. H. (2015). Corporate social responsibility and media coverage. *Journal of Banking & Finance, 59*, 409–422.
Coombs, W. T. (2007). *Ongoing crisis communication: Planning, managing, and responding* (2nd ed.).
Eble, M. (2013). Medienmarken im Social Web: Wettbewerbsstrategien und Leistungsindikatoren von Online-Medien aus medienökonomischer Perspektive, .
Elia, C., Russ-Mohl, S. (2007). Vermittler und Richter in heikler Position. In Neue Zürcher Zeitung, Oct. 5. EJO-Posting of Oct. 11. Vermittler und Richter in heikler Position | Europäisches Journalismus-Observatorium (EJO) | (ejo-online.eu).
Fehrle, B., Höges, C., Weigel, S. (2019). In eigener Sache. Der Fall Relotius. Abschlussbericht der Aufklärungskommission. In SPIEGEL, Nr. 22, May 25, 2019), (pp. 130–146). https://www.SPIEGEL.de/media/media-44564.pdf.
Ferrucci, P. (2019). The end of ombudsmen? 21st-century journalism and reader representatives. *Journalism & Mass Communication Quarterly, 96*(1), 288–307.
Fichtner, U. (2018, 19 Dezember). Manipulation durch Reporter. SPIEGEL legt Betrugsfall im eigenen Haus offen. https://www.SPIEGEL.de/kultur/gesellschaft/fall-claas-relotius-SPIEGEL-legt-betrug-im-eigenen-haus-offen-a-1244579.html.
Fröhlich, R. (1994). Rundfunk-PR im Kontext. Historische und organisationstheoretische Bedingungen am Beispiel ARD. Studien zur Sozialwissenschaft, Opladen.
Funck, A. (2015). Die transparente Redaktion: ein Ansatz für die strategische Öffentlichkeitsarbeit von Zeitungsverlagen. Wiesbaden.
Garcia-Sanchez, I. M., Cuadrado-Ballesteros, B., & Sepulveda, C. (2014). Does media pressure moderate CSR disclosures by external directors? *Management Decision, 52*(6), 1014–1045.
Gündling, U. (2007). Die Neuausrichtung des Zeitungsmarketings durch Customer Relationship Management, Baden-Baden.
Hass, T., Klusmann, S. (2019). Liebe Leserin, lieber Leser. In Fehrle B, Höges C, Weigel S (eds). In eigener Sache. Der Fall Relotius. Abschlussbericht der Aufklärungskommission. In SPIEGEL Nr. 22, May 25, 130. https://www.SPIEGEL.de/media/media-44564.pdf.
Hoffmann, C. P., Russ-Mohl, S. (2020). Der Fall Relotius und die Auswirkungen auf den Spiegel. In Wiske, J. (Hrsg.), Krisenkommunikation komplex: 11 Analysen prominenter Fälle mit medialer Einordnung und Nachbetrachtung beteiligter Experten, Köln, (pp. 210–239).
Holtzhausen, S. (2015). Crossmediale Markenführung. Eine qualitative Studie zur Bedeutung der Markenpositionierung bei Medienunternehmen, Münster.
Klusmann, S., Kurbjuweit, D. (2018, 20 Dezember): Der Fall Relotius. Wir haben sehr viele Fragen an uns selbst. https://www.SPIEGEL.de/politik/deutschland/der-fall-claas-relotius-wir-haben-sehr-viele-fragen-an-uns-selbst-a-1244196.html
Laskin, A. V., & Samoilenko, S. A. (2014). The investor communication strategies of newspaper corporations: A computerized content analysis. *International Journal of Strategic Communication, 8*(3), 196–214.
Lungmus, M. (2019, März). Diese Medien haben Relotius-Texte veröffentlicht. Journalist Nr. 3/2019, (pp. 40–43).
Meier, K. (2019): Quality in journalism. In T. P. Vos/F. Hanusch/D. Dimitrakopoulu/M. Gerertsema-Sligh A. Sehl (eds.), The international encyclopedia of journalism studies (pp. 1–8). Wiley-Blackwell.

Okrent, D. (2008). Public editor #1: The collected columns (with reflections, reconsiderations, and even a few retractions) of the first ombudsman of the New York Times, New York.

Russ-Mohl, S. (1994). Der I-Faktor. Qualitätssicherung im amerikanischen Journalismus—Modell für Europa, Osnabrück/Zürich.

Russ-Mohl, S. (2017). Die informierte Gesellschaft und ihre Feinde. Warum die Digitalisierung unsere Demokratie gefährdet, Köln.

Strömbäck, J. (2005). In search of a standard: Four models of democracy and their normative implications for journalism. *Journalism Studies, 6*(3), 331–345.

Töpfer, A. (2006). Krisenkommunikation: Anforderungen an den Dialog mit Stakeholdern in Ausnahmesituationen. In Schmid, B. F., Lyczek, B. (Hrsg.), Unternehmenskommunikation. Kommunikationsmanagement aus Sicht der Unternehmensführung Wiesbaden, (pp. 351–398).

Trommershausen, A. (2011). Corporate responsibility in Medienunternehmen, Köln.

Weiss-Blatt, N. (2021). The techlash and tech crisis communication. Emerald Group Publishing.

Zerfaß, A. (2014). Unternehmenskommunikation und Kommunikationsmanagement: Strategie, Management und Controlling. In: Zerfaß, A., Piwinger, M. (Hrsg.), Handbuch Unternehmenskommunikation: Strategie—Management—Wertschöpfung (2 ed., pp. 21–80), Wiesbaden.

Zyglidopoulos, S. C., Georgiadis, A. P., Carroll, C. E., & Siegel, D. S. (2012). Does media attention drive corporate social responsibility? *Journal of Business Research, 65*(11), 1622–1627.

# Corporate Social Responsibility: Hiring Requisition in Media Companies?

**Gerrit Boehncke**

## 1 Introduction: The Role of CSR in Top Talent Recruiting

The modern communicative competition confronts companies with numerous challenges. It is not enough to be perceptible, memorable, recognizable, and differentiating as long as the trust between core target groups and the sending brand is missing. Today, the importance of trust in the company and brand is central to long-term and loyal relationships of any kind, whether B2B, B2C, or B2E. Against this background, CSR communication is no longer communicative freestyle but "business imperative" (Schäfer et al., 2020) and assumes central importance for corporate success and thus within strategic corporate communication (Trommerhausen & Karmasin, 2015). Sustainable communication is not only the goal of responsibility management but already a processual expression of the perception of responsibility (Karmasin & Weder, 2008). The responsibility for such *integrated CSR communication* lies with corporate management or the communication strategy.

As a communicative compulsory element, CSR can be justified from several perspectives, like a social level (Metag et al., 2016; Hagedorn, 2019), an economical level (Grabowski, 2019; EY, 2020; Murguia & Lence, 2015; Bocken, 2015), a political level (BMBF, 2020), or a legal level (BGBl. I p. 2152).

While the focus of this book is on CSR communication and its importance for the perception of twin responsibility of media companies at the organizational level, focusing on both business-to-business (B2B) and business-to-consumer (B2C) relationships, this chapter discusses the challenges of twin responsibility for strategic corporate development at the employee level. Employees are an important resource for corporate success and are fiercely contested in the ongoing "war for talent"

G. Boehncke (✉)
Frankfurt am Main, Germany
e-mail: geboehncke@edu.aau.at

 
F. Weder et al. (eds.), *CSR Communication in the Media*, CSR, Sustainability, Ethics & Governance, https://doi.org/10.1007/978-3-031-18976-0_7

(Chambers et al., 1998). In order to win this critical resource, the holistic and lasting communication of a positive employer brand (*Employer Branding*) has become a standard in HR marketing. Recruiting communication, which is aimed at filling specific vacancies, benefits greatly from a positive employer brand. In the course of this chapter, both the communicative efforts from the more holistic employer branding and the more pointed recruiting communication will be considered. Employer branding as the totality of measures to strengthen the employer brand includes a wider range of stakeholders than recruiting communication. While recruiting communication takes place more in formats such as (online) job postings and job ads, employer branding also includes measures such as employee magazines, content marketing in social (Instagram and others) or business web (LinkedIn). Both communication formats are particularly relevant in the media industry, as job applicants and future employees have a particularly strong interaction and broad prior knowledge of marketing formats and channel functions due to their field of work.

Since corporate social responsibility refers to a company's norm- and value-based obligation to the public (Carroll, 1979), it becomes an expression of corporate self-image—the corporate culture. This describes "a pattern of common basic premises (...) for dealing with problems" (Schein, 1985) and central attitude issues. The managers of a company play an increasingly important role in answering questions critical to success and actively shaping corporate culture (Ansoff, 1965; von Alemann, 2015).

Assuming that highly sought-after young executives, as high potentials, can choose between several options when selecting an employer and do not only pay attention to hard factors such as pay (Cable & Judge, 1994; Waples & Brachle, 2020; Twenge & Kasser, 2013), the article analyzes to what extent media companies use CSR as part of their top talent recruiting of potential young executives to increase employer attractiveness and secure talent retention. To this end, the article analyzes the digital recruiting communication of the 10 media companies with the widest reach in Austria and Germany on their web and (social) business network presences. In a first step, a structuring content analysis (Mayring, 1991) is used to examine the significance of CSR content for corporate communication on the business web. Subsequently, the article analyzes the playout and marketing of the career and landing pages using an SEO analysis. This SEO analysis should make it possible to identify the focus topics, keywords, and search intentions of the target group expected by the company.

The article focuses on the career opportunities that are most relevant for junior executives. Both top management traineeships and assistant programs are considered.

This chapter complements existing research on the importance of CSR for overall employer attractiveness (Carlini et al., 2019) by adding a perspective on the narrow—but critical for corporate success—target group of top talents.

Furthermore, by analyzing the status quo of the existing digital communication, the article enables an approach to the strategic expectations of tendering companies and their communication managers. The chapter focuses on the question of whether companies expect CSR awareness and interest among their prospective executives. Does the *business imperative of* CSR communication lead to an imperative for the recruiting of junior executives as well?

## 2 CSR as Awareness Content

In order to be able to adopt a practice-oriented perspective on the use of CSR in the recruiting context and to be able to carry out a content-related deep dive in the following, it must first be explained at which point in the communication between companies and candidates CSR can be of particular relevance. For this purpose, the article follows the "6-Phase Model" of the Candidate Journey by Verhoeven (2016) (see Fig. 1)—which is familiar to many employees in the media industry too.

The candidate journey is based on the customer journey and the AIDA model, which describes the development of a customer in four phases (awareness—interest—desire—action). In the context of modern, digital communication, both journey models consist of various individual points of contact between the company and the individual (customer/candidate), which do not necessarily run linearly. The particular advantage of this model is its long-term nature. The Candidate Journey describes the development of a candidate from the point of approach and enthusiasm for a company (attraction—recruitment), through the specific application phase (application—selection), to hiring and integration (onboarding—retention). Thus, it not only puts into perspective the active talent approach up to the application campaign or hiring but goes beyond that and also includes onboarding and retention. In these two phases, employees experience the previously advertised and certified corporate culture as part of the company for the first time and evaluate it from a new—a company-internal—perspective. It is, therefore, logical to discuss not only

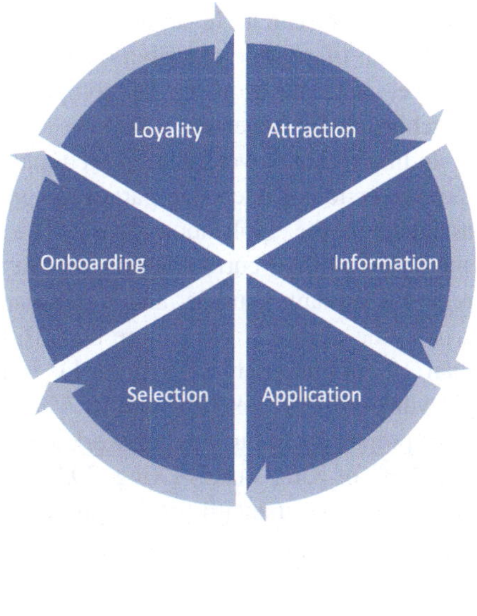


**Fig. 1** Candidate funnel according to Verhoeven (2016)

the approach process but also the evaluation process and possible interactions within the framework of an expanded recruiting concept and the "6-Phase Model."

Since CSR communication has a positive impact on Organizational Attraction (OI) (Saurombe et al., 2017; Schäfer et al., 2020), this chapter focuses particularly on the early phases of "attraction" and "information," within which potential employees should be convinced of the attractiveness of the employer.

Companies have a variety of analogue and digital communication channels and individual media at their disposal for addressing potential applicants in the attraction phase and for presenting the company in the information phase (Dahl et al., 2018; Nied et al., 2018; Wozney et al., 2017; Belinda et al., 2018). An extended selection of individual media is shown in Fig. 2.

Among the large number of touchpoints listed, two channels, in particular, are of particular importance for the target group of young executives, which can affect both the *attraction* and *information* phases. These are career websites and the business networks LinkedIn and Xing.

As business networks, Xing and LinkedIn offer companies the opportunity to promote their corporate culture through different content formats (videos, testimonial statements, linked podcasts, written contributions) both statically in their owned media profiles (Bonchek, 2014) and actively and with clear target group targeting (paid media) (Koch et al., 2018; Liu-Thompkins, 2019). These platforms thus enable both targeted attraction campaigns on the part of the company, which according to Verhoeven are to be located in the attraction phase, as well as temporally sovereign information research by candidates, which can be located within the information phase.

What is special for the business networks is that the content created can be effective at different funnel levels. If it is posted organically, i.e. without any additional monetarily acquired reach, only the users of LinkedIn and Xing who already follow the posting company or have subscribed to the company's content receive the posts. In the following, as with all social networks, the post is placed in the company's timeline. Here, it can be viewed by all interested parties who are conducting research. In addition, however, there is also the possibility of paid promotion of organic contributions and posts and thus to address users who have not yet paid attention to the company and its content (Patrutiu Baltes, 2015; Hutchins, 2016; Vinerean, 2017).

In line with the overarching functionality of business networks, they also support the conversion of interested parties to applicants via Web2Lead form, which can be assigned to the application phase. The relevance of business networks is also supported by their current strong growth in membership. In the so-called D-A-CH region (Germany, Austria, and Switzerland), which is particularly relevant for this article, the networks have more than 16 million users (LinkedIn, 2021) and more than 19 million users (Xing, 2021), respectively, and thus show a membership growth of 45% (LinkedIn) and 27.5% (Xing) within the last three years. Since a large proportion of business network users have management responsibilities—more than 45% of Xing's users (Xing, 2021)—the networks can also be evaluated as the communicative standard for this group of employees.

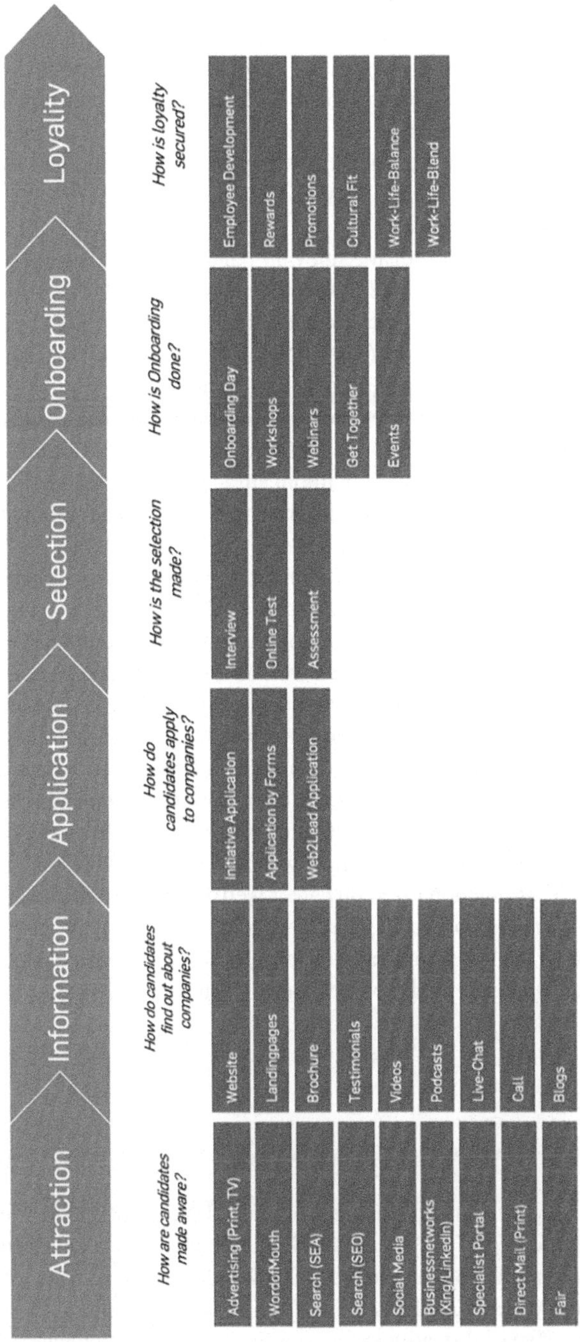


**Fig. 2** The touchpoints of the candidate journey. Own illustration

In addition to business networks, which are central to the interpersonal networking and career communication of prospective executives (McCabe, 2017; Davis et al., 2020), the websites and career sections of companies are particularly important. These generally represent the greatest significance for graduates looking for a job (Neder & Scheller, 2019). With a focus on the target group of junior executives examined here, development programs (such as management traineeships) or qualifying positions (executive assistantships, speaker positions) that are explicitly provided and advertised for top talents and junior executives on specific landing pages are of particular relevance.

Search engine optimization of the relevant pages ensures that the websites and their landing pages are optimized for specific search requests from top talents. A distinction is made between organic content optimization (SEO) (Enge et al., 2012) and paid advertising optimization (SEA) (Schikora, 2014). Within specific career websites, the conversion of a prospect to an applicant is often possible through form applications or is supported by a redirect to an exclusive applicant area. While the specific search engine rankings through SEO/SEA measures are still in competition with external competitors and aim at an initial attraction of the searchers, the website and landing page contents serve the information phase. The following analysis discusses different possible applications of CSR communication within the recruiting funnel in a practice-oriented manner, conducts a cross-channel analysis of the current implementation status, and mirrors this with the expectations of the talent collected via interview. By focusing on the extremely competitive resource of top talents and prospective executives, as well as the simultaneous consideration of the previous channel activities and expectations of the recipients, the article takes a pointed perspective and raises current benchmarks for the area of top talent hiring.

After pointing out CSR as a relevant component of corporate culture and thus the task of executives and describing the particular importance of business networks and career websites/landing pages for addressing and securing the attention of young executives, the following analysis will now provide information on the extent to which CSR content is functionalized on career websites and landing pages as well as on the company profiles of business networks for top talent programs. For this purpose, the career websites and landing pages, as well as the business network presences of the 10 largest media brands from Austria and Germany, are analyzed below.

## 3 Analysis

For the analysis, the five largest privately organized media brands in Germany and Austria were considered. The "brand size" was defined as revenue strength—not reach strength or number of employees (DerStandard, 2020). This criterion was applied because explicitly addressing and training top talent to become junior executives requires special qualification programs (e.g., management traineeships). Such programs and positions mean an additional investment in training the talent,

**Table 1** Overview of the analyzed media brands. Own illustration

| Media brand | Xing profiles | LinkedIn profiles | Country | Focus |
|---|---|---|---|---|
| Red Bull Media House GmbH | Red Bull Media House GmbH | Red Bull Media House | Austria | Graduates |
| Mediaprint Zeitungs- und Zeitschriftenverlag GmbH & Co. KG | Mediaprint Zeitungs- und Zeitschriftenverlag Ges.m.b.H. & Co KG | Mediaprint Zeitungs- und Zeitschriftenverlag Ges.m.b.H. & Co KG | Austria | Unspecific |
| Sytria Media Group AG | Sytria Media Group AG | Styria Media Group AG | Austria | Unspecific |
| ProSiebenSat.1 Puls4 GmbH | ProSiebenSat.1 Puls4 GmbH | ProSiebenSat.1 Puls4 GmbH | Austria | Unspecific |
| Sky Österreich Fernsehen GmbH | Sky Germany GmbH | Sky Germany GmbH | Austria | Young professionals/ executives |
| Bertelsmann SE & Co. KGaA | Bertelsmann SE & Co. KGaA | Bertelsmann SE & Co. KGaA | Germany | Management programs |
| ProSiebenSat.1 Media SE | ProSiebenSat.1 Media SE | ProSiebenSat.1 Media SE | Germany | Graduate programs |
| Axel Springer SE | Axel Springer SE | Axel Springer SE | Germany | Junior executive program |
| Hubert Burda Media Holding Limited Partnership | Hubert Burda Media | Hubert Burda Media | Germany | Management trainee |
| Georg von Holtzbrinck GmbH & Co. KG | Holtzbrinck Publishing Group | Holtzbrinck Publishing Group | Germany | Assistance to the management |

which also means a special financial outlay for the media brands. Also, for this reason, communication was considered at the top corporate level and not at the divisional or individual brand level.

An overview of the media companies studied, their web and business network presences, and their candidate focus is presented in Table 1.

## 3.1 Business Web Content Analysis

Since it is possible for the business networks Xing and LinkedIn to use the existing content as a paid advertising campaign to gain attention, and thus also to place it in the early attention phase (Verhoeven, 2016), the analysis in the following first step looks at the Xing and LinkedIn content of the 10 largest media brands. All the media brands considered have company profiles on both business networks. However, an analysis of the post frequency[1] reveals fundamental differences in the use of the networks.

Both Austrian and German media companies generally published more on LinkedIn ($n = 1446$ posts/year) than on Xing ($n = 801$ posts/year) in the last 12 months. However, the difference for German media companies (Xing = 790 vs. LinkedIn = 1046) was much smaller than for Austrian players (Xing = 11 vs. LinkedIn = 400).

The post frequency of the actively communicating companies also differs greatly. While Bertelsmann is active on Xing almost daily, with an average of 28.5 posts per month, all Austrian media companies post less than 1 post per month on Xing. Of the Austrian media brands, only Sky has more than 50 total posts on Xing.

On LinkedIn, the publishing groups Axel Springer and Georg von Holtzbrinck set the quantitative benchmark with 18.6 posts/month. The German companies are all above 15 posts/month. In Austria, ProSiebenSat.1Puls4 is the quantitative leader with 11.8 posts/month. Styria Media (9.4), Red Bull Media House (6.6), and Mediaprint (4.1) are below 10 posts/month, and Sky Austria has only 1.4 posts/month. For LinkedIn, all companies except Sky (17) and Media print (49) have more than 50 posts for the past year.

Comparability of CSR relevance within the networks is made difficult by the widely varying post frequency as well as the number of total contents on the networks. To nevertheless ensure comparability, the last 50 posts of the media companies were analyzed using a structuring content analysis according to Mayring (1991).[2] Due to the exclusion of the five Austrian media companies for Xing, $n = 750$ posts (Xing = 250; LinkedIn = 500) were ultimately considered for this analysis. Differentiation of the CSR content was made according to the practice-oriented meta-categories "People," "Planet," and "Profit" (Bergmans, 2006). The contributions were assigned using a predefined code catalogue[3], the code basis of which goes back to Dahlsrud's CSR review and definition (Dahlsrud, 2008). The

[1] All Xing and LinkedIn posts on the company profiles from March 13, 2020–March 13, 2021 were evaluated.

[2] Since none of the five Austrian media brands has the required number of Xing posts in the last 12 months, they were excluded from the Xing analysis. For the German media brands, the period under review was extended for ProSiebenSat.1 Media SE compared to the other brands. The 50 most recent Xing posts went back to Sept. 20, 2019 for ProSiebenSat.1 Media SE. Since all media companies could fundamentally show >50 posts on LinkedIn, the observation period for the 50 most recent posts was extended here for the Austrian media companies Sky and Media print. Due to the abstract LinkedIn dating of older posts, however, no time frame can be specified to the day.

[3] People: Society, Inclusion, Diversity, Education, Learning, Equality, Women, PayGap, Health, Life, Psychical Health, E-Health, Mental Health, Poverty, Children, Culture, Antisemitism, Freedom, Queer, LGBTQ+, Policy, Learn, Hate speech, Bullying, People, Nutrition, Tolerance, Together, Tafel ("Tafel" is the only term not translated as it is the German name of a charity) Planet: Environment, Clean, Climate, Climate Change, Climate Crisis, Pollution, Sustainability, FridaysForFuture, Emissions, Water, Tree, Green. Profit: [Employee + Code from People|Planet Category] Talent, Work–Life Balance, Home Office, Remote Office, Candidate, Onboarding, Data Security, Security, Code of Conduct.

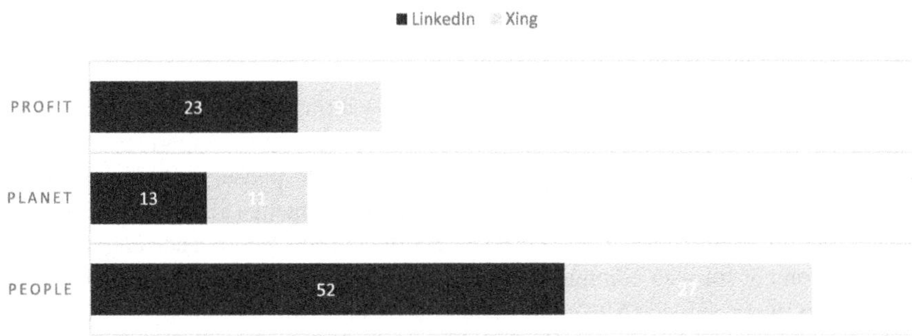


**Fig. 3** Total number of business network posts per CSR dimension. Own illustration

identified code keywords were always used in English and German, as both languages are used in the business profiles. An expansion of the code categories took place, analogous to Dahlsrud's procedure, by adding current CSR-related Google search trends. In addition to explicit terms of corporate commitment, topic terms were also integrated with which companies are concerned or for which countermeasures were presented (#antisemitism, #poverty). Only the immediate post-descriptions (ad title, subtitle, content text, hashtags) were subjected to the structuring content analysis. Linked content or video content was not analyzed, as it would only have a secondary effect on the recipient in the context of an attention campaign.

### *3.2 CSR Content on LinkedIn*

Of the $n = 750$ posts analyzed, 125 had CSR references. The majority of 79 items belonged to the "People" dimension, while 32 items could be assigned to the "Profit" dimension and were thus related to employees or directly to the company. The "Planet" dimension was addressed in only 24 contributions (Fig. 3).

For the individual dimensions, there were also conspicuously dominant keywords. In the *People dimension,* $n = 13$ posts were identified that dealt with providing for and securing the future of children and young people (#Children). The *Planet* dimension was dominated by "Climate*" keywords dominated with $n = 9$, while in the dimension *Profit* equality topics (#Inclusion, #Equality, #Woman) were even more present with $n = 9$ than health topics $n = 4$—despite Corona.

A further discussion of the results is held together with the findings from the web analysis.

### 3.3 Career-Landing Page Content Analysis

As current trend studies (Neder & Scheller, 2019) show, career websites are particularly relevant for the candidate journey. They are the most important source of information for searching users in the information phase, but can also be functionalized to attract attention through targeted content (SEO) or paid (SEA) search engine optimization (Fig. 2). To assess the current significance and potential of CSR content for web communication, the career and landing pages of the media companies were subjected to a two-stage analysis. As a first step, a structuring content analysis according to Mayring was again used to illuminate the potential of CSR aspects within the career pages. Since visitors to a website can only find out about what is explicitly written, just as Google can only find and use for ranking the content that is indexed (Kelsey, 2017), the content analysis of the on-site page text enables initial relevant insights into the potential of CSR aspects. For this purpose, the on-site page text was captured in full using *seorch.de*[4] before the existing CSR code catalogue (see footnote 3) was applied to it.

In the subsequent second analysis step, the actual strategic[5] keyword playout was examined. It shows which of the potential contents are used as attention markers. This second analysis step thus enables a reflection of the page alignment with (expected) search needs and core interests of the target group.

### 3.4 Status Quo of CSR On-Site

The total analyzed on-site text corpus of the 10 career websites included $n = 6786$ words. When applying the CSR code catalogue, codes hit 49 times. Since some of the identified codes were directly related to other defining words (e.g., "*responsibility* towards fellow human beings"), these word references were also evaluated as "CSR content." Thus, CSR contents finally applied to extended word-groups of $n = 76$ (1.11% of the text set).

It is striking in the on-site text analysis that topics of general social responsibility on career websites lag significantly behind the level of the business networks. While contributions on the dimension of *people* still dominated within the business networks, the analysis of the on-site text showed that the absolute majority of CSR codes here relate directly to added value for employees. Only 10 (=20.4%) of the 49 codes identified addressed CSR aspects on a meta-level as CSR or regarding the *planet* or *profit dimension*.

The final keyword analysis was to identify whether some of the CSR terms are also used for explicit SEO optimization. The analysis of the program microsites for

[4]seorch.com: Link: https://seorch.de/ (retrieved 03/15/2021).

[5]Using the SEO analysis tools SEMrush: Online Source: https://de.semrush.com/ (retrieved on 16.03.2021)

junior executives already showed that these are not designed as the first level of information for searching talents. SEMrush identified only $n = 17$ keywords for the microsites considered. None of the keywords was covered by a CSR code.

For the non-specific Austrian career websites, the keyword analysis resulted in $n = 852$ keywords. Again, none of the keywords had a CSR code in them.[6] In an expanding analysis step, the higher-level career sections of the German media brands were also examined. For them, $n = 3984$ keywords were identified by SEMrush. In $n = 10$ cases, a CSR code was hit.

It was striking that only one keyword, "diversity (and inclusion)," was considered for the CSR dimension *people,* while the other keywords were related to *profit.*

# 4 Discussion and Outlook

The individual analysis steps found concrete answers to the central questions of the article, which will be briefly presented and discussed below.

## 4.1 Differences in Website Content and Performance

As a first step, the article asked whether a uniform standard can be identified for media companies in attracting top talent.

The study revealed that there is no international, mechanical standard for recruiting junior executives in media companies. The first look at the career websites already highlighted key differences in recruiting communication. While the German media companies have established explicit microsites for junior executives within their career areas, comparable programs do not exist in Austria. Accordingly, there is also no separate communication for junior executives within the career websites. As an alternative to the explicit training of junior executives in separate programs, two other recruitment options are also used in German media companies, since the few participants in junior executive programs cannot make up the entire executive level of a media company in purely numerical terms: either executives are recruited as already experienced professionals or they are identified at a later stage among operational employees (Thomas, 2016). Future research, as an extension to Leadership Characteristics (Sturm et al., 2017), could examine the extent to which leadership styles differ among "employees explicitly trained as executives," "operationally trained employees," and "recruited experienced executives."

There are also striking differences in performance at the technical level. German media companies communicate more and faster. The recruiting communication of

[6]The pure addressing of talents, applicants, candidates, and employees was excluded and only included if a CSR-specific extension took place.

Austrian and German media companies differs both in terms of content frequency within the business networks and concerning general web optimization. In terms of web performance, the analyzed career pages of the German media companies were already 16-speed points faster for desktop[7]and a whole 24 points for mobile[8] . Comparing the business web profiles, it was noticeable that German media companies generally publish significantly more posts than Austrian media companies. In addition, it was particularly noticeable that Xing plays virtually no role in the content communication of Austrian media companies, while the German companies also communicate regularly here. An interesting question for future research is therefore an international comparison of content communication in terms of frequency and content focus on social networks, which can also be used for recruitment (Khullar & Pandey, 2014; Headworth, 2015; Mičík & Mičudová, 2018).

### *4.2 Relevance of the CSR Dimensions on Corporate Websites and Social Media*

The second question of the article examined which CSR content focuses set on the networks LinkedIn and Xing, as well as the company pages.

Media companies do not focus their CSR communication on ecological aspects. Both the analysis of the business web content and the content analysis of the on-site texts on the career pages showed that the CSR communication of media companies focuses particularly rarely on the *planet* dimension. While the *people* dimension dominated the business web, the direct reference to the *profit* dimension was particularly strong on the career pages. The analysis is therefore in line with the findings of Dahlsrud (2008), which, however, predate the #FridaysforFuture initiative by a long way. Belinda et al. (2018) had still demonstrated a particular attractiveness of environmental CSR in the course of their study. This proven qualitative relevance did not result in a quantitative dominance of environmental CSR for the period under study. Future research could look at the period after the Corona pandemic to see whether the relevance of environmental CSR identified by Belinda et al. leads to a quantitative increase in contributions with an environmental focus, or whether the interest of applicants in this topic declines in the long term.

### *4.3 SEO Strategies of Recruiting Sites*

Finally, the article questioned to the extent there are implications for the SEO alignment of recruiting websites. Although the positive impact on perceived

[7]88/100 German average vs. 72/100 Austrian average. Surveyed with Seorch.de.

[8]76/100 German average vs. 52/100 Austrian average. Surveyed with Seorch.de.

company attractiveness has been repeatedly documented (Lim & Greenwood, 2017; Carlini et al., 2019; Schäfer et al., 2020), only about 1% of the on-site texts on career websites addressed CSR aspects. Even in SEO strategy, the <0.5% of keywords showing a CSR reference do not indicate that communication managers of media companies assign strategic relevance to the topic when recruiting junior executives for touchpoints on the web. Explicitly addressing junior executives poses a particular challenge in this regard. Junior executives and top talents represent a numerically very small target group, which also creates a numerically small search volume and therefore does not become explicitly tangible for SEO analyses. However, the recommendations of tools for SEO performance analysis primarily focus on quantitative indicators and serve to quantitatively ensure success. They identify search volumes, conversion rates, and action rates, which, however, complicates qualitative traffic decisions.

### *4.4 Summary*

Media companies in Germany and Austria have so far used CSR content in different ways. However, CSR content does not (yet) play a really significant or even central role in recruiting and employer branding communication at any company. In particular, the integration into career websites is quantitatively very small. This is understandable in view of the low search volume and the small target group of top talents with an ethical focus, but it raises the question of whether HR and marketing/communications departments in the media companies are already sufficiently well networked to ensure an optimal approach to top talents. In any case, CSR communication has so far only been used to a very limited extent in recruiting communication (of top talent).

## 5 Conclusion and Directions for Future Research

The analysis of business web content showed that CSR aspects are not only a *business imperative* for B2C but are also used for corporate-candidate relationships as well as B2B relationships, but so far more on business networks than within landing pages for high potential recruiting. The low level of integration in the on-site text as well as the low significance of CSR for the keyword strategy of career websites contradict the idea that media companies specifically want to interest their junior executives in the company through CSR communication on the explicit program websites.

The signs indicate that the research question posed, whether media companies try to convince their young executives through CSR communication within career websites, has to be answered rather negatively.

Although CSR sensitivity thus represents a relevant competence for corporate success, it seems not explicitly addressed within the career websites as a decision criterion, area of interest, and character trait (Belinda et al., 2018). However, it remains unclear whether this strategic measure results from a generalized Big Data primacy, or whether successfully recruited junior executives were also captured as lead personas in a Small Data approach (Kitchin & Lauriault, 2015) and influenced the strategic direction of recruiting communications. A reflection of these findings and the question of their generalizability needs to be addressed by future research. A possible answer could be provided by qualitative interviews with HR and communication managers on the company side, as well as applicants and candidates on the job search on the talent side. However, the target group of high potentials considered in this article as future managers is difficult to grasp for empirical studies due to their hardly predictable career paths.

In addition to this central question, however, there are also some limitations to the existing analysis that could be resolved in the course of further research. In parts these solutions will require a strongly modified data access or a change in the legal data utilization, as well as technical data accessibility.

A general limitation draws on a fundamental observation. The Austrian and German media brands differ fundamentally in the entry programs they offer and the entry positions they provide for highly qualified talent and potential junior executives. While all of the five largest German media brands advertise explicit programs for junior executives and provide information about them, among the Austrian brands only the special case Sky offers a corresponding program—which is also currently suspended. However, this is a conscious decision on the part of the media companies. Other Austrian companies, such as Coca Cola HBC Austria and Marsh and McLennan, do offer explicit programs for young executives.

This fact is relevant for the analysis design insofar as it does not allow an *explicit* targeting of high potentials for the Austrian media market to be examined. Instead, it will be particularly important to consider whether the web and SEO strategies of Austrian companies that see high potentials as only one possible target group for their website differ fundamentally from the web and SEO strategies for explicit high potentials.

Another limitation is the limited number of countries, companies, and hierarchical levels that were compared in this study. To gain a complete picture of the significance of CSR communication in recruiting communication, additional countries, industries, and also media companies and corporate levels should be examined for their communication.

As well the timeframe of 2020 and 2021 can be seen as a limitation. The period under consideration was strongly influenced by the Corona pandemic, which also had an impact on recruiting and recruiting communications. In addition, the Corona pandemic was also responsible for the reduction or suspension of programs and positions for the promotion and development of young talent, even at major media companies. Furthermore, it is reasonable to hypothesize that in the wake of the pandemic, there has been a shift in CSR focus to social and specific health-related issues. Future research could compare the pre-Covid, in-Covid, and post-Covid recruiting content communication. This kind of content analysis is of interest to researchers in the fields of media and communication studies, as well as human resource marketing.

A last limitation also exists in the SEO analysis or mechanical-technical analyzability. For example, a comparative view of the SEA campaigns of the individual media companies was not possible. The long-term and international analysis of the business network campaigns is also not yet technically possible and would require support and data released by the companies themselves. For this, future research needs explicit access to corporate communications managers and data experts. This approach seems particularly accessible and valuable to researchers in online marketing and data science.

Now that this chapter has looked very much at the external communication of CSR content in the course of recruiting and employer branding communication, future research could, however, also take a look at the internal communication of (media) companies.

An analysis of the internal communication and coordination between human resources managers and recruiters on the one hand, and editors and online marketing managers on the other, could therefore be of particular interest for future research projects on recruiting young executives. The hypothesis that editors and online marketing managers want to ensure quantitative web success and therefore focus recruiting communication content on more superficial brand and benefits topics that are of particular interest to the broad majority of applicants (Saks et al., 1996; Simpson & Aprim, 2018) would be supported by the results obtained from the on-site text analysis and the keyword analysis. A consideration of this hypothesis seems to be an interesting field of research especially for researchers of Management studies, but also for researchers of sociology and psychology.

## References

Ansoff, H. I. (1965). *Corporate strategy: An analytical approach to business policy for growth and expansion*. McGraw-Hill.

Belinda, C. D., Westermann, J. W., & Bergmann, S. M. (2018). Recruiting with ethics in an online era: Integrating corporate social responsibility with social media to predict organizational attractiveness. *Journal of Vocational Behavior, 109*, 101–117.

Bergmans, F. (2006). Integrating people, planet and profit. In J. Jonker & M. de Witte (Eds.), *Management models for corporate social responsibility* (pp. 117–125). Springer.

Bocken, N. M. P. (2015). Sustainable venture capital—catalyst for sustainable start-up success? *Journal of Cleaner Production, 108*, 647–658.

Bonchek, M. (2014). Making sense of owned media. *Harvard Business Review*. https://hbr.org/2014/10/making-sense-of-owned-media. Accessed 27 May 2021

Cable, D. M., & Judge, T. A. (1994). Pay preferences and job search decisions: A person-organization fit perspective. *Personnel Psychology, 47*, 317–348.

Carlini, J., Grace, D., France, C., & Lo Iacono, J. (2019). The corporate social responsibility (CSR) employer brand process: An integrative review and comprehensive model. *Journal of Marketing Management, 35*, 182–205.
Carroll, A. B. (1979). A three-dimensional conceptual model of corporate performance. *The Academy of Management Review, 4*, 497–505.
Chambers, E., Foulon, G., Handfield-Jones, M., Henkin, H., Steven, M., & Michaels, E. G. (1998). The war for talent. *The McKinsey Quarterly, 3*, 44–57.
Dahl, A. J., D'Alessandro, A. M., Peltier, J. W., & Swan, E. L. (2018). Differential effects of omnichannel touchpoints and digital behaviours on digital natives' social cause engagement. *Journal of Research in Interactive Marketing, 12*, 258–273.
Dahlsrud, A. (2008). How corporate social responsibility is defined: An analysis of 37 definitions. *Corporate Social and Environmental Management, 15*, 1–13.
Davis, J., Wolff, H. G., Forret, M. L., & Sullivan, S. E. (2020). Networking via linked in: An examination of usage and career benefits. *Journal of Vocational Behavior*. https://doi.org/10.1016/j.jvb.2020.103396
Enge, E., Spencer, S., Fishkin, R., & Stricchiola, J. (2012). *The art of SEO*. O'Reilly Media.
Hagedorn, G. (2019). FridaysForFuture: Digitization makes protest more scientific. *Earth System Knowledge Platform*. www.eskp.de. https://doi.org/10.2312/eskp.016
Headworth, A. (2015). *Social media recruitment: How to successfully integrate social media into a recruitment strategy*. Kogan Page Publishers.
Hutchins, A. L. (2016). Beyond resumes: LinkedIn for marketing educators. *Journal of Research in Interactive Marketing, 10*, 137–147.
Karmasin, M., & Weder, F. (2008). Corporate communicative responsibility corporate communication as responsibility management. *prmagazine, 5*, 57–62.
Kelsey, T. 2017. SEO basics. In *Introduction to search engine optimization*. : Apress.
Khullar, A., & Pandey, P. (2014). Effective use of social media recruiting. *International Journal of Intercultural Information Management, 4*, 216–227.
Kitchin, R., & Lauriault, T. P. (2015). Small data in the era of big data. *Geo Journal, 80*, 463–475.
Koch, T., Gerber, C., & DeKlerk, J. (2018). The impact of social media on recruitment: Are you LinkedIn? *SA Journal of Human Resource Management, 16*. https://doi.org/10.4102/sajhrm.v16i0.861
Lim, J. S., & Greenwood, C. A. (2017). Communicating corporate social responsibility (CSR): Stakeholder responsiveness and engagement strategy to achieve CSR goals. *Public Relations Review, 43*, 768–776.
Liu-Thompkins, Y. (2019). A decade of online advertising research: What we learned and what we need to know. *Journal of Advertising, 48*, 1–13.
Mayring, P. (1991). Qualitative Inhaltsanalyse. In U. Flick, E. V. Kardoff, H. Keupp, L. V. Rosenstiel, & S. Wolff (Eds.), *Handbuch qualitative Forschung: Grundlagen, Konzepte, Methoden und Anwendungen* (pp. 209–213). Beltz.
McCabe, M. B. (2017). Social media marketing strategies for career advancement: An analysis of LinkedIn. *Journal of Business and Behavioral Sciences, 29*, 85–99.
Metag, J. M. S., Schäfer, T., Füchslin, T. B., & Kleinen-von-Königslow, K. (2016). Perceptions of climate change imagery: Evoked salience and self-efficacy in Germany, Switzerland, and Austria. *Science Communication, 38*, 197–227.
Mičík, M., & Mičudová, K. (2018). Employer brand building: Using social media and career websites to attract generation Y. *Economics and Sociology, 11*, 171–189.
Murguia, J. M., & Lence, S. H. (2015). investors' reaction to environmental performance: A global perspective of the Newsweek's 'Green Rankings'. *Environmental and Resource Economics, 60*, 583–605.
Neder, Pablo and Scheller, Stefan. 2019. Digital candidate journey. https://docs.google.com/viewerng/viewer?url=https://persoblogger.de/download/digital-candidate-journey-studie-2019-2020-zum-recruiting-von-studenten-studierenden/?ind%3D1604916837734%26filename%3

DStudie_Digital_Candidate_Journey_2019_2020_final.pdf%26wpdmdl%3D11374%26refresh%3D60486c437df621615359043%26open%3D1. Accessed 27 May 2021.
Nied, J., Bilger, R., & Ferdinand, H. M. (2018). Candidate experience: Eine empirische Analyse der Recruiting-Prozesse am Beispiel der Generation Y. *Markenbrand, 7*, 35–46.
Patrutiu Baltes, L. (2015). Content marketing—the fundamental tool of digital marketing. *Series V Economic Sciences, 8*, 111–118.
Saks, A., Wiesner, W., & Summers, R. (1996). Effects of job previews and compensation policy on applicant attraction and job choice. *Journal of Vocational Behavior, 49*, 68–85.
Saurombe, M., Barkhuizen, E. N., & Schutte, N. E. (2017). Management perceptions of a higher educational brand for the attraction of talented academic staff. *SA Journal of Human Resource Management, 15*, 1–10.
Schein, E. H. (1985). *Organizational culture and leadership*. Jossey-Bass Publishers.
Schäfer, S. D., Terlutter, R., & Diehl, S. (2020). Talking about CSR matters: Employees perception of and reaction to their company's CSR communication in four different CSR domains. *International Journal of Advertising, 39*, 191–212.
Schikora, C. (2014). *Basics online marketing-SEO, SEA and social media*. Herbert Utz Verlag.
Simpson, S. N. Y., & Aprim, E.-K. (2018). Do corporate social responsibility practices of firms attract prospective employees? Perception of university students from a developing country. *International Journal of Corporate Social Responsibility, 3*, 1–11.
Sturm, R. E., Vera, D., & Crossan, M. (2017). The entanglement of leader character and leader competence and its impact on performance. *The Leadership Quarterly, 28*, 349–366.
Thomas, M. (2016). *Internes Headhunting: Talente entdecken–Führungskräfte entwickeln*. Springer VS.
Trommershausen, A., & Karmasin, M. (2015). Corporate social responsibility in media companies. In J. Krone & T. Pellegrini (Eds.), *Handbook of media economics. Springer Nachschlage Wissen* (pp. 1–34). Springer VS.
Twenge, J. M., & Kasser, T. (2013). Generational changes in materialism and work centrality, 1976-2007: Associations with temporal changes in societal insecurity and materialistic role modelling. *Personality and Social Psychology Bulletin, 39*, 883–897.
Verhoeven, T. (2016). *Candidate experience. Approaches for a positive experienced employer brand in the application process and beyond*. Springer VS.
Vinerean, S. (2017). Content marketing strategy. Definition, objectives and tactics. *Expert Journal of Marketing, 5*, 92–98.
von Alemann, A. (2015). *Gesellschaftliche Verantwortung und ökonomische Handlungslogik: Deutungsmuster von Führungskräften der deutschen Wirtschaft*. Springer VS.
Waples, C. J., & Brachle, B. J. (2020). Recruiting millennials: Exploring the impact of CSR involvement and pay signalling on organizational attractiveness. *Corporate Social Responsibility and Environmental Management, 27*, 870–880.
Wozney, L., Turner, K., Rose-Davis, B., & McGrath, P. J. (2017). Facebook ads to the rescue? Recruiting a hard-to-reach population into an Internet-based behavioural health intervention trial. *Internet Interventions*. https://doi.org/10.1016/j.invent.2019.100246

## *Online Sources*

BMBF. (2020). Press release 145/2020. https://www.bmbf.de/de/karliczek-wir-wollen-europa-bis-2050-zum-ersten-klimaneutralen-kontinent-machen-12687.html. Accessed 27 May 2021.
DerStandard. (2020). Österreichs größte Medienhäuser 2020—und ihre Erwartungen in Zeiten von Corona. https://www.derstandard.at/story/2000118737432/oesterreichs-groesste-medienhaeuser-2020-und-ihre-erwartungen-in-zeiten-von. Accessed 27 May 2021.

EY. (2020). Sustainable consumption. https://assets.ey.com/content/dam/ey-sites/ey-com/de_de/news/2020/05/ey-nachhaltiger-konsum-2020.pdf. Accessed 27 May 2021.

Grabowski, M. (2019). *"Nachhaltiges Leben 2020"—Acht von zehn Befragten wollen das kommende Jahr nachhaltig gestalten*. Spiegel Media. https://www.spiegelgruppe.de/news/pressemitteilungen/detail/nachhaltiges-leben-2020-acht-von-zehn-befragten-wollen-das-kommende-jahr-nachhaltig-gestalten. Accessed 25 May 2021

LinkedIn. (2021). LinkedIn's economic graph. https://economicgraph.linkedin.com/. Accessed 27 May 2021.

Xing. (2021). Facts and figures. https://werben.xing.com/daten-und-fakten#:~:text=As%20f%C3%BChrendes%20professional%20business%2DNetwork,unique%20reach%20in%20highly%20relevant%20targetgroups. Accessed 27 May 2021.

# Smart Exclusion: How May Digital Platforms Hinder Inclusivity within News Organizations?

**Cinzia Dal Zotto, Afshin Omidi, and Esmaeil Norouzi**

## 1 Introduction

The nature of work in the news media industry has changed dramatically over the recent decades. This can be significantly attributed to the rise of digital platforms, which have drastically changed the core processes within newsrooms as well as the workforce skills they require, the methods of data collection, processing, and analysis, and how news content is distributed (Lischka, 2021). The rising popularity of digital platforms among the news audience has also completely changed how the audience can interact with the media (Napoli, 2016). These developments have led to the dawn of a new order in the news ecosystem, exemplified by the emergence of new phenomena such as big data (S. C. Lewis & Westlund, 2015).

According to some indicators, the use of digital platforms is growing at an annual rate of 25% (Kässi & Lehdonvirta, 2018). To cope with these changes, news organizations have increasingly turned to digital intermediaries to carry out different activities, influencing the nature of work in these organizations (Graham et al., 2017). With the rise of global Internet usage and access, people worldwide now generate gigantic amounts of data every day. To operate optimally in the world of digital media, many news organizations analyze these big data with analytics software such as *Chartbeat* in order to identify the favorite topics of their audiences, determine how content can be conveyed more effectively, analyze essential

C. D. Zotto (✉) · A. Omidi
Institute of Management, Faculty of Economics and Business, University of Neuchâtel, Neuchâtel, Switzerland
e-mail: cinzia.dalzotto@unine.ch; afshin.omidi@unine.ch

E. Norouzi
Faculty of Management, University of Teheran, Tehran, Iran
e-mail: norouziesmaeil@ut.ac.ir

 
F. Weder et al. (eds.), *CSR Communication in the Media*, CSR, Sustainability, Ethics & Governance, https://doi.org/10.1007/978-3-031-18976-0_8

concepts, and better understand the general trends of their profession (Dwyer & Martin, 2017).

The effects of digital platforms on organizations in general, and news media in particular, can be, though not exhaustively, discussed in the light of several dimensions. The first dimension is how these platforms have transformed the cost structures of information dissemination by allowing individuals and organizations to publish their information for free, a phenomenon that was somehow impossible before the arrival of digital technologies (Parker et al., 2016). The second dimension is how digital platforms have facilitated communications between organizations and individuals, enabling them to interact with each other at any time and from any place (Brynjolfsson & Smith, 2000; Roshandel Arbatani et al., 2019). This allows journalists to be constantly present on the platforms, working directly or indirectly for news organizations, a trend that has implicitly intensified their work pressures. The third dimension is the marketing and advertising opportunities offered by digital platforms and other digital tools, which enable news organizations to communicate more effectively with specific segments of their audience and leverage this capability to target their offer better and increase their revenue (Chua & Westlund, 2019). Thus, to compensate for the declining revenues from traditional sources, news organizations proactively seek for additional revenue sources through and within digital spheres (Murschetz et al., 2020). Because of these transformations, many news media now expect journalists to be familiar with digital advertising and marketing methods and use this knowledge to produce digital revenue-generating content (Petre, 2018).

Finally, the fourth and most critical dimension is how digital platforms have challenged the functioning logic of news organizations (Andreassen et al., 2018). In the past, news media were characterized by a physical infrastructure where newsrooms with their news production and dissemination systems were allocated, and journalists gathered. In other words, employees working in the media sector were obliged to attend specific physical places to gather and process information, conduct their analyses and present the outcome to the editors. The rise of digital platforms has eliminated this location dependency, changing the way information is collected, how it is analyzed, as well as the schedule of activities (Dal Zotto & Omidi, 2020).

Drawing on the data colonialism perspective and the labor process theory, this chapter conceptually and critically argues how digital platforms may hinder news organizations' inclusivity and thus social responsibility if their in-depth implications at the macro and micro levels are not well understood. While the study recognizes the positive potentialities of digital technologies for fostering media activities and workplaces, it also shows a capitalist mode of production steering them, which may have some critical consequences for the management of news media. More clearly, digital technologies could be harnessed to extract, produce, analyze, and distribute data intelligently and thus allow news organizations to reach both profit and social impact. However, these technologies are transforming the nature of news work, whose concern is primarily building an informed and democratic society. By creating a new, uncontested reality for defining the workplace in news organizations, smart technologies are directly influencing the recruitment and selection process within news organizations, which are now mainly, if not only, looking for those talents who appear to be ideally best suited to work within this new digital context. As a result, a process that we call "smart exclusion" emerges, which impacts on news organizations' social license to operate. This paper tries to conceptualize the process of smart exclusion in news media, suggesting adopting a symbiotic perspective for dealing with the related challenges and ensuring socially responsible behavior of news organizations.

For this purpose, the present chapter has been organized as follows. First, we explain the theoretical approach upon which our ideas have been developed. Second, we discuss the building blocks residing in our main theories, namely labor process and data colonialism, through which we have analyzed the phenomenon as mentioned above. Third, the concept of smart exclusion has been clarified into three, though related, thematic strands, including data, skills, and control strategies. Finally, we present our solutions toward inclusive news organizations in the age of digital platforms.

## 2 Theoretical Development Approach

This section describes the philosophical approach of the research and explains how it addresses the problems discussed in the introduction. While using data colonialism and labor process theories, which are pretty critical, we acknowledge that new technologies have had undeniable positive impacts in facilitating news media activities, expanding the range of analyses, and improving fact-checking tasks by journalists (Chan-Olmsted, 2019). However, as long as the capitalist mode of production plays a fundamental role in how media organizations operate, digitalization can also have detrimental consequences for the professional autonomy of news workers and the future of well-informed democratic societies. To explore this subject, we conducted a systematic search for the existing evidence concerning the implications of the rise of digital platforms. We tried to integrate and interpret such implications based on the aforementioned theoretical perspectives: labor process and data colonialism. The former helped us understand managerial and organizational changes inside media firms. At the same time, the latter enabled us to grasp some fundamental changes in the external environment of media organizations, all brought about by the domination of data in our everyday lives.

To design our conceptual inquiry, we have adjusted our research based on the guidelines introduced by Jaakkola (2020) and Hulland (2020) and benefited from the provoking theorization suggested by Sandberg and Alvesson (2021). The provoking theory is an approach that helps us challenge existing mindsets about a specific phenomenon and proposes a different way of thinking based on the dialectics between existing theories on one side and unknowns on the other side. While the effects of a data-driven environment on journalism seem to indicate the dawn of a new, promising era in this profession, there could be some new effects being away from our sights. Considering that smart and digital technologies are frequently applauded as promising forces to create inclusive workplace environments in different industries (Holford, 2019; Järventie-Thesleff et al., 2014; Malmelin & Villi, 2017; Oldham & Da Silva, 2015), we will argue how these new technologies may have some other threatening effects on the very nature of news work, leading to a situation where a smart exclusion process could be explored.

# 3 Theoretical Background

## *3.1 Data Colonialism*

The ever-increasing pervasiveness of social media and digital technologies has ushered in the age of big data, in which massive amounts of information about our everyday life, which are collected through our smartphones or computers, can be used for different purposes in our personal and social lives. This age is also characterized by the remarkable capability of information systems to collect and analyze vast amounts of information about human performance and preferences. Such information tends to be organized in databases that are optimized for its processing and analysis (Kalantzis-Cope, 2016). Within this context, a new form of capitalism has emerged, which, in today's information-centered world, has been labeled as data colonialism or big data colonialism (J. E. Cohen, 2018; Srnicek, 2017; Zuboff, 2015). Industrialization made capitalism the primary regulator of many societal affairs in the industrial age. Today, big data has opened the door to a more advanced type of capitalism whose purpose is still primarily to raise capital. However, unlike its industrial age predecessor, which involved acquiring and possessing wealth, property, factories, and so on, this new form of capitalism seeks to collect and possess the data that people produce on digital platforms every day.

Data colonialism involves turning new social relations, i.e., data relations that produce raw inputs for information processing, into a resource for generating new forms of economic value. The value of extraction processes depends on the comprehensiveness of the data produced. As Schneier (2015, pp. 8–27) points out, before the Internet, data sources about people's social life were limited to corporates' customer records, direct marketing information, insurance credit records, and public records through which organizations were able to provide higher quality goods and services to meet consumer demands better. Today, advanced processing machines extract data from our everyday lives more thoroughly than could ever be achieved in any previous social arrangements (Couldry & Mejias, 2019).

The main actors of data colonialism are the companies that collect information about people's social and daily activities and translate them into quantitative data for analysis and ultimately profit generation. While companies can use various software and hardware for this purpose, the essential instruments of data colonialism in today's world are ever-increasing digital platforms. According to Couldry and Mejias (2019), people interact with each other and different organizations in various ways. In data colonialism, platforms define these interactions as data relations of individuals to produce input for their information analysis needs. Since social platforms store people's social activities and communications as quantitative data, they can repeat this social "datafication" process as frequently as they need. Calzati (2020) also states that platforms transform the qualitative relations of their users into quantitative data, which algorithms can analyze for value generation. According to Arvidsson (2016), the main goal of these digital platforms is to encourage people to be more active and express their inner thoughts. As with the very nature of capitalism, data colonialists encourage data production on platforms to have continuously more data for analysis. Their purpose is to access and analyze the data generated by social users. As Fuchs (2017) points out, promoting a particular pattern of behavior involving the voluntary sharing of information and loyalty to the platform can significantly facilitate the data mining process.

## 3.2 Labor Process Theory

In the Marxist school, labor is essential for changing the environment and developing human capacities and potentials (Marx, 2004). According to Marx, labor is a combination of conception and execution, in which the conception part distinguishes humans from animals (Fuchs, 2014a). Following the development of Marx's conceptualization of labor process and Baran and Sweezy's (1966) theorization of monopoly capital, Braverman (1998 [1974]) published the book "Labor and Monopoly Capital," in which he challenged the emerging management discourses that began primarily with Taylor's scientific methods, showing that the concept of labor is being degraded. In the following years, this book received much attention as one of the essential works in the Marxist's view of labor and employment.

During its inception, the labor process theory was mainly focused on industrial factories, the nature of which used to depend on several groups of factors, including the physical workspace where the idea becomes a commodity, particular groups of people such as managers and workers, and a series of rules and regulations set to facilitate the work. In the labor process theory, this workplace is called the point of production (Braverman, 1998; Burawoy, 1982; Thompson & Smith, 2009). Nevertheless, this definition has lost some of its validity in advanced capitalist societies, where the points of production are no longer fixed in nature and are dependent on place, time, and people (Thompson & Smith, 2009). For example, unlike industrial factories, digital platforms do not require any physical environment and include a virtual space where employees work and communicate. Even the relationship between employer and employee on these platforms could be different. For example, research in the field of the service economy and sharing economy (Bolton, 2004; Sturdy et al., 2001; Vincent, 2011), as well as creative work (McKinlay & Smith, 2009) provided evidence that platforms tend to determine relations between employees and employers as if they would play the role of policymakers.

One of the significant ideas of the labor process theory receiving substantial attention from organizational sociologists (e.g., Edwards, 1979; Friedman, 1977) is managerial control. Littler (1990) states that, unlike in the pre-industrial era when managers had relatively limited power in controlling their workers, industrialization helped modern management invent new labor control strategies in line with the central goals of capitalism, namely to increase the profitability and stability of operations. In the modern era, digital technologies have become practical tools for controlling and monitoring the workplace (Beverungen et al., 2015; Moore, 2017; Moore & Robinson, 2016). Even though constantly evolving to meet the emerging needs, such new control mechanisms operate in two ways: through coercion and/or by manufacturing consent in the workplace. Burawoy (1982) states that while workers think they are resisting new ways of managing, in the end, they participate in generating added value for employers. Hochschild (1983) further specified that managerial control focuses on manipulating workers' minds and emotions rather than directing employees through physical means.

The second central concept of the labor process theory is the "deskilling process," which is closely related to managerial control. This concept refers to the division of labor into more superficial elements to help senior managers monitor work processes (Smith, 1994). Frequently, organizations need employees with two groups of skills, cognitive and executive. Since managers often seek to increase efficiency while decreasing employment costs, they pay more attention to executive skills like technical and operational competencies (T. Lewis, 2007). In recent years, some scholars have acknowledged that the labor process theory (LPT) research has not progressed well enough to address digital age-related issues, like the expansion of digital platforms and the consequent changes in the required skills (Gandini, 2019). The most important criticism in this area is that emerging phenomena such as platform economics have led to the formation of a new work culture, which has not received enough attention from LPT scholars (Böhm & Land, 2012, p. 220; Cappelli & Keller, 2013). Nevertheless, we believe that this theory can offer valuable insights into the effects of digital platforms on the required skills within media organizations.

## 4 Smart Exclusion: How Do Digital Platforms Hinder Inclusivity?

This section explains how digital platforms are implicitly changing the professional practices inside news organizations under the guise of rationalization of media labor within a volatile and technology-driven environment. There exist, we firmly believe, no destructive forces in the very nature of digital platforms, but the managerial and political relations driving the application of these technologies in news media may lead to some detrimental effects, both for journalists in particular and democratic societies more in general. Thus, we show how a data-driven environment, at a macro level, has been forcing news organizations to transform their organizational practices in order to survive and prosper significantly. Then, we illustrate how the required journalism skills are being changed as a response to the new realities of the digitized media environment, thereby creating new criteria for selecting candidates for the journalistic profession. Finally, we discuss how digital platforms have been harnessed to control media labor in a capitalistic spirit and how this can threaten journalism's future and news organizations' public service mission. We attempt, by and large, to depict a picture in which a process of what might be called "smart exclusion" may emerge. In this context, we conceive "smart exclusion" as a process characterized by three interconnected trends: (1) a data-driven media ecosystem that is considerably influencing the way news organizations operate, (2) the resulting increased technological component in the skills required for journalists to become professionals in the field, and finally, in the worst case, (3) the exclusion of technologically unskilled journalists, even though highly committed and driven by the core values of journalism for acquiring newsworthy information and disseminating it to the public to contribute to a democratic society.

## *4.1 Data as a Key Driver in Changing the News Profession*

This section argues how a data-driven media ecosystem has had a considerable impact on how journalistic work is evolving. It intends to open up a discussion on the ways performance and success of news organizations in the media markets are highly interconnected with their data-driven abilities. More clearly, we show how the importance of data in media management is congruent with the insights offered by the data colonialism perspective.

Over the past two decades, the rise of digital platforms has led to the proliferation of data-driven news journalism. During this time, many data analysis tools have been designed and used in newsrooms and editorial offices to help journalists make sense of big data in their work processes, including analyzing and interpreting trending events. Previous studies have shown that leading news media companies are very interested in applying analytics tools within newsrooms (Petre, 2018; Tandoc, 2014), and indeed analytics tools have been used throughout various media, including television and newspapers. For example, such tools helped to rank television programs in terms of audience attraction (Napoli, 2011), and allowed newspapers to assess audience tastes and set newsroom's priorities accordingly to increase their circulation. Today, data analytics enables the monitoring of users' activities on various social networks: media can grasp how audiences reach news feeds and what they think about it, learn more about news traffic trends, content viewing time, how many times the content is clicked, and shared (Hendrickx et al., 2021; Petre, 2018). This emerging trend is seemingly common in advertising and commercial organizations, where data tracking aims to identify demand and adjust products and services for higher profits.

With the rising importance of and attention to data production, many journalists have decided to become active players in big data generation and take up the task of increasing the traffic of news websites (Anderson, 2011; Dreyfus & Dreyfus, 2005). This is not inherently destructive as journalists can create data traffic by producing content according to the values and principles of their profession rather than merely serving platforms' algorithms to get more audience attention. Journalists, who ideally should produce news content with a high degree of independence, are now though obliged to grapple with technological issues and their effects on the very nature of their profession (Bunce, 2019). Consciously understanding such effects is thus of critical importance for journalists.

Nowadays, some journalists produce content that is more likely to become a trend rather than addressing relevant public issues, which implicitly makes them a servant of algorithms and big data generation systems (N. S. Cohen, 2019; Lischka, 2021). This could potentially be a significant threat to the values of journalism, and in the worst-case scenario, can lead to a future in which the core values of journalism in terms of informing the public are "smartly excluded" and wholly replaced with algorithmic values (Gillespie, 2020). Nieborg and Poell (2018) refer to this as "platformization," defining it as the process in which journalists are increasingly directed and organized through the platform ecosystems. Also, since platforms are constantly changing their algorithms based on continuous incoming evaluations, journalists who want to keep producing trending content must become increasingly tech-savvy and work harder to understand the rules of this technological game.

## *4.2 Changing Skills Requirements*

While the last section paid attention to the role of data in driving the operations of news organizations, we here indicate how those trends could have altered the very skills journalists need to be successful from the employers' point of view. While we do not condemn the changes brought about by a data-driven media environment in news work, we argue how those emerging skills become the foundations upon which news organizations strategize their core functionalities, including defining the essential skills for recruiting the proper journalists. The arrival of computers and the information age forced journalists to improve their technical skills and encouraged them to become familiar with new technologies such as different digital platforms and analytics. With the growing use of smartphones, journalists had to learn a form of journalism tailored to this new space and produce content customized for these devices. The outcome of the interaction between news media and smartphone technologies was the emergence of a new strand called mobile journalism (Agur, 2019; Pignard-Cheynel & Van Dievoet, 2019). As the name implies, mobile journalism is a new form of journalism in which journalists use the growing capabilities of mobile phones or notebooks with Internet access to collect, edit, and distribute news from the community around them (Kumar & Mohamed Haneef, 2018).

Social networks and platforms have become a place to produce and distribute news content. This development has forced content producers to keep up with technological advancements and adjust their work activities accordingly to attract and retain users' attention from various devices (Crespo et al., 2020). Thus, instead of improving their journalistic knowledge, newsroom workers constantly need to acquire new platform-dependent capabilities. Indeed, with the proliferation of online news sites and the emergence of social networks as channels for high-speed distribution of news content, audience fragmentation has increasingly intensified, and competition in attracting audience attention has become fierce. Consequently, the advertising revenues of news organizations decreased (Murschetz, 2020). To counter these threats, news organizations have mainly reacted with cost-cutting measures, progressively reducing the number of employees in their newsrooms and focusing on recruiting people with a wide array of skills in digital content creation and distribution (N. S. Cohen, 2015b; Kumar & Mohamed Haneef, 2018; Petre, 2018).

According to Nygren (2014), employing highly skilled staff decreases the number of employees and improves organizational efficiency in the newsroom. However, this raises pressure on journalists to adapt to new technologies (N. S. Cohen, 2015a). Nowadays, many journalists either perform micro-repetitive tasks requiring specific technological capabilities or are assigned more general tasks not requiring any particular skill (Silberman & Irani, 2016). News workers often play the role of a technical operator and attempt to develop the capabilities of a programmer or data analyst. The digital ecosystem has led news organizations to be more interested in hiring technically skilled journalists, leaving other people with valuable skills but less knowledgeable in technology out of the loop and thus "smartly excluded" from selection processes.

## *4.3 Changing Control Strategies*

Data analysis software offers new ways of managing and supervising work (N. S. Cohen, 2019). For instance, journalistic activities can be standardized and become more logical, trackable, and reliable in revenue generation. The performance of published content can be easily and precisely checked through quantitative monitoring tools such as news analytics dashboards. Media work processes appear thus somewhat comparable to those set in place by the twentieth-century scientific management method (Braverman, 1998; Taylor, 1911). Such metric-driven rationalization of journalistic labor processes has brought journalists to resist and react to this news business approach, mostly based on analytic tools (Bunce, 2019; Christin, 2017). From an employer's point of view, it leads to a higher level of managerial control over journalistic work, greater efficiency, and profits. From an employee's point of view, this means lower levels of independence for journalists (Carlson, 2017). By supplying new controlling tools, technology gives managers more power to control over the allocation of financial and nonfinancial resources. This may ultimately lead to hiring, promoting, and paying journalists based on their capabilities to generate traffic revenues and thus boost the overall organizational performance.

In practice, digital technologies are also used to create ways to control labor processes in different workplaces (Moore, 2017; Moore & Robinson, 2016). Digital control strategies can be used to continuously monitor and evaluate employee performance and productivity (Gandini, 2019; Rosenblat & Stark, 2016). In news organizations that use newsroom dashboards, journalists are required to generate a certain amount of content each day based on assigned tasks. Dashboards are then used to control journalists' performance based on the amount of feedback their content receives on digital platforms in the form of likes, shares, comments, and audience engagement (N. S. Cohen, 2016, 2019). The more successful a journalist is in this context, the more income she/he will earn and the more secure her/his job will be in the organization. These criteria reflect the effect of managerial decisions, particularly performance evaluations, monitoring, and control on media work. The digitization of activities often more clearly defines tasks and thus helps employees better understand and perform their duties. However, if used by managers as an instrument to control and supervise work, digitization can elicit an adverse reaction from the employees' side. Indeed, journalists need an open and innovative working environment to thrive (Hesmondhalgh & Baker, 2013). If constantly threatened by technology-enabled supervision, they may feel limited in their independence and creativity and even decide to quit their employer (Gandini et al., 2016; Leung, 2014). We can thus say that dashboard-driven news organizations may provoke an unintentional "smart exclusion" of talented and socially responsible journalists, thereby putting in danger their public service mission.

## 5 Toward Inclusive News Media: A Symbiotic Perspective

While digital platforms have created an excellent opportunity for news organizations to engage with their audience in previously impossible ways, it needs to be well appreciated that news media, especially public media, are essentially meant to raise public awareness and stimulate open public discourse in society (Downey & Fenton, 2003; Fuchs, 2014b). Letting managerial decisions such as the recruitment and selection of employees be solely driven by the imperatives of digital strategies may mislead media organizations from their public service mission. We argue that, in order for news media to represent and include the diversity of public opinions and thus facilitate open discourse, they should rather assess job candidates based on their journalistic values and skills. The eventually missing digital skills can be the objective of human resource development plans. That said, media managers need to ask themselves a critical and fundamental question: what is the main objective of a news organization? If the answer is to maximize news traffics and increase audience engagement on social media, then only the most tech-savvy applicants are most probably going to be considered for entering news organizations. On the contrary, if news media care more about their public service mission than data analytics, committed applicants who care about journalistic principles and the values of democratic societies will not be excluded in favor of people who are just skilled in digital technologies. In this case, media managers should rather think about how new technologies could be harnessed to enhance their social responsibility and augment the quality of journalistic content.

One of the big fears in journalism is how technology may gain a central role in defining the priorities, processes, and practices within news media organizations in favor of a capitalist mode of production. In the following, we discuss this issue and highlight choices in dealing with the emerging challenges. As humans' ability to process and predict phenomena is primarily limited by their cognitive capacities, digital technologies are increasingly being harnessed to accomplish diverse tasks and complete humans' limitations (Brynjolfsson & McAfee, 2014). As Jarrahi (2018) points out, humans and technology are complementary. Artificial intelligence can, for instance, extend human cognition when addressing complexity. However, through their holistic and intuitive approach, humans can better guide organizational decision-making in conditions of uncertainty. This means that technology should be designed to augment and not substitute human contributions. In the context of news media, it is argued that new technologies on one hand increase the workload of journalists, but on the other hand they can also allow them to be faster, more flexible, and even more creative (Kumar & Mohamed Haneef, 2018). Indeed, digital technologies can complement journalistic activities, thereby significantly improving the effectiveness of news media. By including smartphones, cameras, and computers in their newsrooms, media organizations attempt to create a symbiosis between traditional journalism and new technologies. While many believe that journalism with data and technological tools would produce one-dimensional narratives and neglect the human elements, the symbiosis with machines may allow journalists to access narratives that are directly related to people's lives (Chan-Olmsted, 2019). A symbiotic approach may enable journalism to discover human problems and trigger public agendas for developing potential solutions. Ideally, from a symbiosis point of view, rather than seeing digital platforms as crucial determinants of the news media environment, media managers would benefit by making better decisions, all in favor of a well-informed public and thus the public service mission of news media. In this sense, Marconi (2020) talks about iterative journalism, specifying that surrounding human reporters with artificial intelligence can augment their abilities: by allowing journalists to break more news more quickly, artificial intelligence can simultaneously free up their time for deeper analysis. The newsmaker can this way be comforted by having "a new set of tools to uncover the truth, while also knowing that no algorithm will ever take over her journalistic judgement" (Marconi, 2020, p. 156).

Thus, by adopting a symbiotic perspective, media managers could set up an organizational configuration in which technology augments human performance and prevents highly committed journalists from falling into the process of smart exclusion. By no means do we claim that digital skills and ongoing developments in the digital platform ecosystem should be ignored. Instead, we argue that these technologies should be viewed and treated as critical tools for further developing news staff's intellectual skills, decision-making, and professional independence.

After recruiting people with appropriate mental aptitude and talent to carry out quality journalism, media managers might provide them with the required skills that they will need to perform in the digital age through complementary human resource development programs.

## 6 Conclusion

This chapter discussed how digital platforms change the way news media operate, creating a new working environment in which we may observe a process called smart exclusion. We attempted to theoretically examine this emerging phenomenon in the light of three dimensions: (a) the role of data as a key driver in changing the news media profession, (b) the changes in journalism skills requirements, and (c) the transformations in the strategies for controlling media workers. The review of the existing literature showed that while many newsrooms have introduced new technologies in their operations to enable journalists to work faster and with greater flexibility, the rising economic interest in analytics and digital platforms has created a form of journalism that involves an extensive use of these tools in a capitalistic spirit; a trend that other researchers have also observed. In this new ecosystem, media managers should be aware that some very talented and committed journalists may risk being excluded from their profession, jeopardizing the public service mission of news media organizations, as we clarified in the previous sections. New technologies should be viewed as great opportunities to augment human potential and provide new development paths for the newsrooms' staff to deal with these challenges. In this line, a development-oriented human–technology symbiosis could provide a fertile ground to address the ways digital technologies might be harnessed in media work so that the social responsibility dimensions of diversity and inclusion are ensured within news media organizations.

We encourage future researchers to address how journalists, in different contexts, think about and reflect on the skills required in a digitized media environment and how they mentally, emotionally, and behaviorally respond to these shifts. In particular, it could be interesting to understand how media workers may resist the simultaneous deskilling and multi-skilling processes affecting their jobs, which are significantly intensified by advanced digital platforms. Another highly relevant strand of research could be dedicated to defining appropriate criteria for recruiting and selecting future journalists to avoid the potential exclusion of candidates who, regardless of their commitment and inspiration to the very ideals of journalism, lack smart technology abilities.

## References

Agur, C. (2019). Insularized connectedness: Mobile chat applications and news production. *Media and Communication, 7*(1), 179–188. https://doi.org/10.17645/mac.v7i1.1802

Anderson, C. W. (2011). Between creative and quantified audiences: Web metrics and changing patterns of newswork in local US newsrooms. *Journalism, 12*(5), 550–566. https://doi.org/10.1177/1464884911402451

Andreassen, T. W., Lervik-Olsen, L., Snyder, H., Van Riel, A. C. R., Sweeney, J. C., & Van Vaerenbergh, Y. (2018). Business model innovation and value-creation: The triadic way. *Journal of Service Management, 29*(5), 883–906. https://doi.org/10.1108/JOSM-05-2018-0125

Arvidsson, A. (2016). Facebook and finance: On the social logic of the derivative. *Theory, Culture & Society, 33*(6), 3–23. https://doi.org/10.1177/0263276416658104

Baran, P. A., & Sweezy, P. A. (1966). *Monopoly capital*. Monthly Review Press.

Beverungen, A., Böhm, S., & Land, C. (2015). Free labour, social media, management: Challenging Marxist organization studies. *Organization Studies, 36*(4), 473–489. https://doi.org/10.1177/0170840614561568

Böhm, S., & Land, C. (2012). The new 'hidden abode': Reflections on value and labour in the new economy. *The Sociological Review, 60*(2), 217–240. https://doi.org/10.1111/j.1467-954X.2012.02071.x

Bolton, S. (2004). Conceptual confusions: Emotion work as skilled work. In C. Warhurst, I. Grugulis, & E. Keep (Eds.), *The skills that matter* (pp. 19–37). Palgrave Macmillan.

Braverman, H. (1998). *Labor and monopoly capital: The degradation of work in the twentieth century*. NYU Press.

Brynjolfsson, E., & McAfee, A. (2014). *The second machine age: Work, progress, and prosperity in a time of brilliant technologies*. WW Norton & Company.

Brynjolfsson, E., & Smith, M. D. (2000). Frictionless commerce? A comparison of internet and conventional retailers. *Management Science, 46*(4), 563–585.

Bunce, M. (2019). Management and resistance in the digital newsroom. *Journalism, 20*(7), 890–905. https://doi.org/10.1177/1464884916688963

Burawoy, M. (1982). *Manufacturing consent: Changes in the labor process under monopoly capitalism*. University of Chicago Press.

Calzati, S. (2020). Decolonising "data colonialism" propositions for investigating the realpolitik of today's networked ecology. *Television & New Media, 22*, 1–16. https://doi.org/10.1177/1527476420957267

Cappelli, P., & Keller, J. R. (2013). Classifying work in the new economy. *Academy of Management Review, 38*(4), 575–596. https://doi.org/10.5465/amr.2011.0302

Carlson, M. (2017). Automating judgment? Algorithmic judgment, news knowledge, and journalistic professionalism. *New Media & Society, 20*(5), 1755–1772. https://doi.org/10.1177/1461444817706684

Chan-Olmsted, S. M. (2019). A review of artificial intelligence adoptions in the media industry. *International Journal on Media Management, 21*(3–4), 193–215. https://doi.org/10.1080/14241277.2019.1695619

Christin, A. (2017). Algorithms in practice: Comparing web journalism and criminal justice. *Big Data & Society, 4*(2), 1–14. https://doi.org/10.1177/2053951717718855

Chua, S., & Westlund, O. (2019). Audience-centric engagement, collaboration culture and platform counterbalancing: A longitudinal study of ongoing sensemaking of emerging technologies. *Media and Communication, 7*(1), 153. https://doi.org/10.17645/mac.v7i1.1760

Cohen, N. S. (2015a). From pink slips to pink slime: Transforming media labor in a digital age. *The Communication Review, 18*(2), 98–122. https://doi.org/10.1080/10714421.2015.1031996

Cohen, N. S. (2015b). Entrepreneurial journalism and the precarious state of media work. *South Atlantic Quarterly, 114*(3), 513–533. https://doi.org/10.1215/00382876-3130723

Cohen, N. S. (2016). *Writers' rights: Freelance journalism in a digital age*. McGill-Queen's Press-MQUP.

Cohen, J. E. (2018). The biopolitical public domain: The legal construction of the surveillance economy. *Philosophy & Technology, 31*(2), 213–233. https://doi.org/10.1007/s13347-017-0258-2
Cohen, N. S. (2019). At work in the digital newsroom. *Digital Journalism, 7*(5), 571–591. https://doi.org/10.1080/21670811.2017.1419821
Couldry, N., & Mejias, U. A. (2019). Data colonialism: Rethinking big data's relation to the contemporary subject. *Television & New Media, 20*(4), 336–349. https://doi.org/10.1177/1527476418796632
Crespo, M., Pinto-Martinho, A., Foà, C., Paisana, M., & Caldeira Pais, P. (2020). Business models of journalistic Startups in Portugal: An analysis of product innovation, dissemination and monetization in media enterprises. *Nordic Journal of Media Management Issue, 1*(2), 261–296. https://doi.org/10.5278/njmm.2597-0445.5194
Dal Zotto, C., & Omidi, A. (2020). Platformization of media entrepreneurship: A conceptual development. *Nordic Journal of Media Management, 1*(2), 209–233. https://doi.org/10.5278/njmm.2597-0445.5234
Downey, J., & Fenton, N. (2003). New media, counter publicity and the public sphere. *New Media & Society, 5*(2), 185–202. https://doi.org/10.1177/1461444803005002003
Dreyfus, H. L., & Dreyfus, S. E. (2005). Peripheral vision: Expertise in real world contexts. *Organization Studies, 26*(5), 779–792. https://doi.org/10.1177/0170840605053102
Dwyer, T., & Martin, F. (2017). Sharing news online. *Digital Journalism, 5*(8), 1080–1100. https://doi.org/10.1080/21670811.2017.1338527
Edwards, R. (1979). *Contested terrain*. Heinemann.
Friedman, A. (1977). *Industry and labour*. Macmillan.
Fuchs, C. (2014a). *Digital labour and Karl Marx*. Routledge.
Fuchs, C. (2014b). Social media and the public sphere. *TripleC: Communication, Capitalism & Critique. Open Access Journal for a Global Sustainable Information Society, 12*(1), 57–101. https://doi.org/10.31269/triplec.v12i1.552
Fuchs, C. (2017). *Social media: A critical introduction*. Sage.
Gandini, A. (2019). Labour process theory and the gig economy. *Human Relations, 72*(6), 1039–1056. https://doi.org/10.1177/0018726718790002
Gandini, A., Pais, I., & Beraldo, D. (2016). Reputation and trust on online labour markets: The reputation economy of elance. *Work Organisation, Labour and Globalisation, 10*(1), 27–43.
Gillespie, T. (2020). Content moderation, AI, and the question of scale. *Big Data & Society, 7*(2), 205395172094323. https://doi.org/10.1177/2053951720943234
Graham, M., Hjorth, I., & Lehdonvirta, V. (2017). Digital labour and development: Impacts of global digital labour platforms and the gig economy on worker livelihoods. *Transfer: European Review of Labour and Research, 23*(2), 135–162. https://doi.org/10.1177/1024258916687250
Hendrickx, J., Eladio, M., Heritiana, R., & Ballon, P. (2021). Becoming the data-informed newsroom? The promotion of audience metrics in the newsroom and journalists' interactions with them. *Digital Journalism, 9*, 1–16. https://doi.org/10.1080/21670811.2021.1890170
Hesmondhalgh, D., & Baker, S. (2013). *Creative labour: Media work in three cultural industries*. Routledge.
Hochschild, A. R. (1983). *The managed heart*. University of California Press.
Holford, W. D. (2019). The future of human creative knowledge work within the digital economy. *Futures, 105*, 143–154. https://doi.org/10.1016/j.futures.2018.10.002
Hulland, J. (2020). Conceptual review papers: Revisiting existing research to develop and refine theory. *AMS Review, 10*(1–2), 27–35. https://doi.org/10.1007/s13162-020-00168-7
Jaakkola, E. (2020). Designing conceptual articles: Four approaches. *AMS Review, 10*(1–2), 18–26. https://doi.org/10.1007/s13162-020-00161-0
Jarrahi, M. H. (2018). Artificial intelligence and the future of work: Human-AI symbiosis in organizational decision making. *Business Horizons, 61*(4), 577–586. https://doi.org/10.1016/j.bushor.2018.03.007

Järventie-Thesleff, R., Moisander, J., & Villi, M. (2014). The strategic challenge of continuous change in multi-platform media organizations—A strategy-as-practice perspective. *International Journal on Media Management, 16*(3–4), 123–138. https://doi.org/10.1080/14241277.2014.919920

Kalantzis-Cope, P. (2016). Whose data? Problematizing the 'gift' of social labour. *Global Media and Communication, 12*(3), 295–309. https://doi.org/10.1177/1742766516676207

Kässi, O., & Lehdonvirta, V. (2018). Online labour index: Measuring the online gig economy for policy and research. *Technological Forecasting and Social Change, 137*, 241–248. https://doi.org/10.1016/j.techfore.2018.07.056

Kumar, A., & Mohamed Haneef, M. S. (2018). Is Mojo (En)De-skilling? Unfolding the practices of mobile journalism in an Indian newsroom. *Journalism Practice, 12*(10), 1292–1310. https://doi.org/10.1080/17512786.2017.1389291

Leung, M. D. (2014). Dilettante or renaissance person? How the order of job experiences affects hiring in an external labor market. *American Sociological Review, 79*(1), 136–158. https://doi.org/10.1177/0003122413518638

Lewis, T. (2007). Braverman, Foucault and the labor process: Framing the current high-skills debate. *Journal of Education and Work, 20*(5), 397–415. https://doi.org/10.1080/13639080701814315

Lewis, S. C., & Westlund, O. (2015). *Big Data and Journalism. Digital Journalism, 3*(3), 447–466. https://doi.org/10.1080/21670811.2014.976418

Lischka, J. A. (2021). Logics in social media news making: How social media editors marry the Facebook logic with journalistic standards. *Journalism, 22*(2), 430–447. https://doi.org/10.1177/1464884918788472

Littler, C. R. (1990). The labour process debate: A theoretical review 1974–88. In D. Knights & H. Willmott (Eds.), *Labour process theory* (pp. 46–94). The Macmillan Press.

Malmelin, N., & Villi, M. (2017). Media work in change: Understanding the role of media professionals in times of digital transformation and convergence. *Sociology Compass, 11*(7), e12494. https://doi.org/10.1111/soc4.12494

Marconi, F. (2020). *Newsmakers: Artificial intelligence and the future of journalism*. Columbia University Press.

Marx, K. (2004). *Capital: Volume I*. Penguin UK.

McKinlay, A., & Smith, C. (2009). *Creative labour: Working in the creative industries*. Palgrave Macmillan.

Moore, P. V. (2017). *The quantified self in precarity: Work, technology and what counts*. Routledge.

Moore, P., & Robinson, A. (2016). The quantified self: What counts in the neoliberal workplace. *New Media & Society, 18*(11), 2774–2792. https://doi.org/10.1177/1461444815604328

Murschetz, P. C. (2020). State aid for independent news journalism in the public interest? A critical debate of government funding models and principles, the market failure paradigm, and policy efficacy. *Digital Journalism, 8*(6), 720–739. https://doi.org/10.1080/21670811.2020.1732227

Murschetz, P. C., Omidi, A., Oliver, J. J., Kamali Saraji, M., & Javed, S. (2020). Dynamic capabilities in media management research. A literature review. *Journal of Strategy and Management, 13*(2), 278. https://doi.org/10.1108/JSMA-01-2019-0010

Napoli, P. M. (2011). *Audience evolution: New technologies and the transformation of media audiences*. Columbia University Press.

Napoli, P. M. (2016). The audience as product, consumer, and producer in the contemporary media marketplace. In G. F. Lowe & C. Brown (Eds.), *Managing media firms and industries* (pp. 261–275). Springer International Publishing. https://doi.org/10.1007/978-3-319-08515-9_15

Nieborg, D. B., & Poell, T. (2018). The platformization of cultural production: Theorizing the contingent cultural commodity. *New Media & Society, 20*(11), 4275–4292. https://doi.org/10.1177/1461444818769694

Nygren, G. (2014). Multiskilling in the newsroom--de-skilling or re-skilling of journalistic work? *The Journal of Media Innovations, 1*(2), 75–96. https://doi.org/10.5617/jmi.v1i2.876

Oldham, G. R., & Da Silva, N. (2015). The impact of digital technology on the generation and implementation of creative ideas in the workplace. *Computers in Human Behavior, 42*, 5–11. https://doi.org/10.1016/j.chb.2013.10.041

Parker, G. G., Van Alstyne, M. W., & Choudary, S. P. (2016). *Platform revolution: How networked markets are transforming the economy? And how to make them work for you*. WW Norton & Company.

Petre, C. (2018). Engineering consent: How the design and Marketing of Newsroom Analytics Tools Rationalize Journalists' labor. *Digital Journalism, 6*(4), 509–527. https://doi.org/10.1080/21670811.2018.1444998

Pignard-Cheynel, N., & Van Dievoet, L. (2019). *Journalisme mobile: Usages informationnels, stratégies édioriales et pratiques journalistiques*. De Boeck Supérieur.

Rosenblat, A., & Stark, L. (2016). Algorithmic labor and information asymmetries: A case study of Uber's drivers. *International Journal of Communication, 10*, 27.

Roshandel Arbatani, T., Norouzi, E., Omidi, A., & Valero-Pastor, J. M. (2019). Competitive strategies of mobile applications in online taxi services. *International Journal of Emerging Markets, 16*, 113. https://doi.org/10.1108/IJOEM-01-2019 0029

Sandberg, J., & Alvesson, M. (2021). Meanings of theory: Clarifying theory through typification. *Journal of Management Studies, 58*(2), 487–516. https://doi.org/10.1111/joms.12587

Schneier, B. (2015). *Data and goliath: The hidden battles to collect your data and control your world*. WW Norton & Company.

Silberman, M., & Irani, L. (2016). Operating an employer reputation system: Lessons from Turkopticon, 2008-2015. *Comparative Labor Law & Policy Journal, 36*, 1.

Smith, V. (1994). Braverman's legacy. *Work and Occupations, 21*(4), 403–421. https://doi.org/10.1177/0730888494021004004

Srnicek, N. (2017). *Platform capitalism*. John Wiley & Sons.

Sturdy, A., Grugulis, I., & Willmott, H. (2001). *Customer service: Empowerment and entrapment*. Palgrave Macmillan International Higher Education.

Tandoc, E. C. (2014). Journalism is twerking? How web analytics is changing the process of gatekeeping. *New Media & Society, 16*(4), 559–575. https://doi.org/10.1177/1461444814530541

Taylor, F. W. (1911). *The principles of scientific management*. Harpers and Brothers.

Thompson, P., & Smith, C. (2009). Labour power and labour process: Contesting the marginality of the sociology of work. *Sociology, 43*(5), 913–930. https://doi.org/10.1177/0038038509340728

Vincent, S. (2011). The emotional labour process: An essay on the economy of feelings. *Human Relations, 64*(10), 1369–1392. https://doi.org/10.1177/0018726711415131

Zuboff, S. (2015). Big other: Surveillance capitalism and the prospects of an information civilization. *Journal of Information Technology, 30*(1), 75–89. https://doi.org/10.1057/jit.2015.5

# CSR in the News Media Industry in Times of the Climate Crisis: A Critical Reflection

Livia Regen and Matthias Karmasin

## 1 Introduction: CSR in Times of Crises and Calls for System Transformation

The idea that corporations are embedded in society and thus carry responsibilities that go beyond their own profit interests has been discussed for some time now and in depth from a number of theoretical perspectives, including business ethics, stakeholder theory, stewardship theory, institutional theory, corporate citizenship, and corporate social performance (Bashtovaya, 2014; Karmasin & Apfelthaler, 2017; Karmasin & Litschka, 2017). The rise of CSR as a salient topic on the public and corporate mind has been accompanied by a flurry of research across academia as Ragas & Roberts (2009, p. 267) summarize. Today, CSR refers to the totality of a firm's economic, legal, ethical, and philanthropic responsibilities (Carroll, 1991 (1979); Karmasin & Litschka, 2017; Schwartz & Carroll, 2003), including the triple bottom line, volunteer work, assistance to individuals in need, the promotion of health and environmental awareness, and the support of local communities and causes (Freeman & Hasnaoui, 2011). Approaches to CSR range from legislating and mandating to recommending (ibid.).

However, it is central to point out that CSR always has been understood as an alternative *within* capitalist economic relations, and thus does not call into question basic capitalist institutions such as the market, private ownership, and profit. So far, merely the *way in which* profits are made and *how* they are reinvested has been criticized via the normative category of responsibility to stakeholders and society.

L. Regen (✉)
Institute of Technology Assessment, Austrian Academy of Sciences, Wien, Austria
e-mail: livia.regen@oeaw.ac.at

M. Karmasin
CMC Austrian Academy of Science/University of Klagenfurt, Vienna, Austria
e-mail: matthias.karmasin@aau.at



F. Weder et al. (eds.), *CSR Communication in the Media*, CSR, Sustainability, Ethics & Governance, https://doi.org/10.1007/978-3-031-18976-0_9

This also holds true for the media industry where recently CSR as a management tool has become an integrated part of strategies, corporate credos, and mission statements (e.g. Batko & Kreft, 2017; Bracker, 2017; Ingenhoff & Bachmann, 2017; Karmasin et al., 2014; Koinig et al., 2019; Tench et al., 2012; Toker, 2013; Trommershausen & Karmasin, 2020), often combined with the communication of CSR activities (e.g. Gulyás, 2009; Ingenhoff & Kölling, 2012; Kalinowska-Żeleźnik et al., 2017; Karmasin & Bichler, 2017). For the implications of CSR on media corporations operating across cultures, c.f. Karmasin in this volume.

Given the lack or insufficiency of public regulation regarding the climate crisis at the national and international scale, CSR for many has come to be viewed as a means to respond to the several social and ecological challenges *within* the constraints of carbon capitalist economic relations. At the same time, scholars have made the case for a necessary and more profound restructuring of global economies, a so-called social-ecological transformation, in order to adequately respond to multiple intersecting social-ecological crises, thus aiming to overcome the boundaries and pitfalls of the existing economic order emerging from Western industrialized societies (Asara et al., 2015; Brand, 2016; Brand & Wissen, 2017). The question that arises from these two trajectories (i.e. change from within the system vs. change *of* the system) is to what degree CSR can be regarded as an effective means to reduce social and ecological pressures, and to what degree it might distract from, or prevent, regulation that is more in line with social-ecological sustainability and the targets laid out in the Paris Agreement (Paris Agreement, 2015). Given that CSR adapts to and operates within an economic system whose pillars are themselves in their current configuration largely unsustainable, the inherent limitations of CSR in regard to contributing profoundly tackling challenges posed by the climate crisis ought to be critically reflected (Regen, 2021). These, as will be elaborated on further down below, include characteristics of CSR such as the voluntary nature of commitments, the low level of public accountability, and the potential for the appropriation of ethical intents for profit generation.

While these characteristics, arguably, hold true for CSR in any industry, this contribution is particularly concerned with news media organizations as both cultural and economic actors (Karmasin, 2006, p. 292) with their location at the intersection of science, politics, and society (M. T. Boykoff & Yulsman, 2013) and as core agents in the construction of the social reality of crises, including the climate crisis (Corfee-Morlot et al., 2007). In the following, we aim to shine light on the question of to what degree CSR can be regarded as an adequate means for news media to respond to the climate crisis. The contribution draws on literature at the intersection of CSR, the news media and the climate crisis, as well as on empirical work that has been conducted as part of Regen's thesis entitled "Between carbon capitalism, ethics and science: coverage and operational responses to the climate crisis by selected English-language news media" (2021). It both makes the case for an extended conceptualization of CSR in times of the climate crisis based on best practice examples and elaborates on the limitations of social responsibility in the corporate sphere.

## 2 The Double Responsibility of News Media in the Climate Crisis

Given the context of an anthropogenic climate crisis with escalating potential (Oreskes, 2004; Ripple et al., 2019) as part of the multiple crises (Brand, 2009) that are rooted in growth-centred globalized carbon capitalism (Di Muzio, 2015; Malm, 2013; Moore, 2017; Steffen et al., 2015), the news media as reality-constructing entities carry a double responsibility:

*Firstly*, to communicate the climate crisis in accordance with the scientific consensus (Oreskes, 2004), i.e. communicative responsibility.

*Secondly*, to make economic decisions that reduce the news media industry's direct or indirect impact on the global climate, i.e. an ecological responsibility.

The latter is both tied to a conceptualization of the media as economic institutions (as analyzed by scholars with critical political media economy approaches (Fuchs, 2017; Knoche, 2001)) and—given digitalization—media infrastructure that is closely entangled with extractivism, labour exploitation, and hazardous waste disposal (Kannengießer, 2020). This contribution will mainly focus on the news media's communicative responsibility and its ecological responsibility for indirect ecological impacts, i.e. through advertising, but will also briefly touch on direct ecological impacts as they might become part of an updated understanding of CSR for the news media industry in the climate crisis. In the following, we aim, on the one hand, to contribute to an extended understanding of CSR in the news media industry in times of the climate crisis, and, on the other hand, to critically reflect on the limitations of CSR.

As regards newspapers' perceived communicative responsibility, the literature reveals the following: while differences occur across geographical, economic, and political (discursive) contexts, levels of newspaper quality, and newspapers' political orientations (Regen, 2021), there appears to be an overall tendency for the news media's growing interest in reporting on the climate crisis (Schmidt et al., 2013). Consequently, what we call communicative responsibility is slowly becoming integrated into newsrooms, albeit to different degrees and levels of quality. At the same time, the news media can present a space for the reproduction of climate science scepticism or denialism (Brüggemann & Engesser, 2014; Painter & Gavin, 2016). A news medium's communicative and ecological responsibility—while at times portrayed as separate—are in fact neatly intertwined: advertising, as the financial backbone for the news media industry, has been shown to influence reporting choices such as a news medium's reduced likelihood to cover corporate irresponsibility in case of relevant advertising relations with the same corporation (Stäbler & Fischer, 2020). Similarly, perceptions of communicative responsibility might be shaped by the anticipation of advertising opportunities, such as during the launch of new car models, which have been found to influence the tone in which the climate crisis is covered (Beattie, 2020). This serves to illustrate that the perception of public responsibility in the news media industry may not only vary between news outlets, but may also vary within a given outlet over time, based on economic incentives, and thus reflections of the context of growth-powered economies. In a context in which economic considerations tend to outweigh ethical concerns, CSR as voluntary self-regulation carries the danger of being an all-too-easy adaptable tool that fluctuates together with economic pressures and incentives rather than an assurance that the news media industry will act in a manner that aids societies navigate the multiple crises. As Besio and Pronzini (2014) contend, CSR can result in both "workable solutions or merely reinforce the status quo, depending on what [...] corporations make of it" (2014, p. 287).

Besio and Pronzini (2014) identify two camps in the evaluation of the role of CSR in times of the climate crisis; on the one hand, literature points to the danger of CSR tools serving little more than corporate image improvements, without "appl[y]ing [ethics] in the core operations and decisions" (2014, p. 288), thus co-opting ethical reasoning for profit generation purposes (ibid., p. 289). On the other hand, literature is found that centres around the claim that high-profit rates and ethical decision-making can go hand in hand (e.g. Robin, 2016). In recent years, however, a growing bulk of post-growth and degrowth literature has shown that profit-driven economic growth is correlated with increased resource and energy throughput as well as labour exploitation, and thus necessarily entails stark social and ecological implications within and across generations that is globally unevenly distributed (Haberl et al., 2020; Parrique et al., 2019). Given the fact that CSR operates within and does not per se challenge the existing growth- and profit-centred economic system which lies at the heart of the climate crisis, the limitations of CSR in regard to an adequate response to the climate crisis become evident.

## 3 Towards an Extended Understanding of News Media CSR in the Climate Crisis

Before discussing the insights gained in Regen's (2021) thesis that will serve to illustrate which practices have the potential to extend the understanding of possible CSR practices in the news media industry, and point to the limitations of CSR as a tool in the context of the global climate crisis, we will briefly introduce the research method.

### *3.1 A Note on the Research Method*

In order to holistically explore the context-dependent nature of the news media as content, economic entities, and infrastructure with social and ecological implications, an exploratory and descriptive multiple-case study (Baxter & Jack, 2010; Yin, 2018) was conducted as part of Regen's (2021) thesis. The in-depth case study relied on different sources of evidence: firstly, qualitative interviews with representatives of four major news outlets in Australia, India, the UK, and the USA with both editorial and business operations personnel; secondly, a document analysis with editorial and company operational documents (e.g. editorial guidelines, ethics codes, style guides, pledges, company reports, financial statements, and corporate guiding principles); thirdly, a media analysis, including news clippings about the given media outlet as represented by different news organizations. The case selection was performed in an exploratory manner and based on a set of minimum requirements: a) considered part of the "press," b) having a significant international audience or a relatively high influence on one particular country, c) published in English, d) with the company base in an English-speaking country, e) availability of relevant documents of a given media company, f) the readiness by at least one representative of the media company to be interviewed (ideally more, or alternatively: an expert who has conducted research on the given news organization). In addition to the criteria described, a heuristic of political leaning and ownership structure as well as the political context (in particular, the level of correlation between conservativism and climate scepticism) guided the case selection. The following news outlets were eventually studied: The Guardian (as a pioneer regarding responses to the climate crisis, and left-liberal, with a non-profit ownership structure; context of low scepticism); *HuffPost* (formerly publicly traded; large corporation owned; left-leaning; context of high scepticism); *Hindustan Times* (a centrist newspaper, privately owned, context of low scepticism, and development narratives), *The Australian* (right-wing newspaper, context of high scepticism; publicly traded). The data interpretation was guided by cross-case analysis tools (Yin, 2018, 167f).

### 3.2 How Pioneering News Media Understand CSR in the Climate Crisis

The empirical findings, which are based on Regen (2021), reveal the following aspects as central when it comes to news media's attending to their communicative responsibility: The reinterpretation of the notion of journalistic balance; linguistic choices reflecting the extent of the climate crisis; and more holistic approaches to climate crisis coverage.

As regards the reinterpretation of the norm of journalistic balance, scholars have previously pointed to the problem of "balance as bias" (M. Boykoff & Boykoff, 2007; Nerone, 2015, p. 225), rendering scientifically established truths about the climate crisis matters of opinion (McKnight, 2010). A revised approach to journalistic balance that is socially responsible could include a scientifically grounded contextualization of perspectives; in fact, pioneering news media such as *The Guardian* and *HuffPost* do not create balance at an ontological level, but depart from the scientific consensus on the climate crisis and aim to establish balance in regard to the spectrum of potential political responses. In addition, a socially responsible way to communicate the climate crisis could entail a revision of the vocabulary that a given outlet employs to describe the crisis, e.g., by updating journalistic style guides as practised by *The Guardian* and *Hindustan Times*. As the findings show, this could be particularly important in regards to more conventional terminology such as "global warming" and "climate change" vis-à-vis "climate crisis/emergency/breakdown", allowing for framing choices that are aligned with the scientific findings on the crisis' extent. Finally, a refined conceptualization of socially responsible journalism should entail a move from event-based and section-specific coverage towards a more holistic reporting practice that "acknowledges climate change as an overarching phenomenon that touches upon many spheres, notably politics, social life and the economy" (Regen, 2021, p. 86).

As regards ecological responsibility, the CSR understanding for (news) media companies could be extended to include choices related to investments, advertising, and decisions regarding a company's ecological footprint as the empirical findings suggest. *The Guardian* serves as a case in point for a news medium where communicative responsibility and ecological responsibility tend to be relatively well aligned: its owner The Scott Trust launched a medium-term divestment programme from fossil fuels in 2015, which rendered it the globally largest fund to divest from fossils (Carrington, 2015; The Guardian, 2020; The Guardian Media Group, 2020). The divestment strategy went hand in hand with reinvestment in managers with social-ecological criteria and the *Keep it in the Ground* campaign, asking major charitable organizations to follow suit (ibid.). Efforts have additionally been undertaken in regard to advertising, albeit in an incomplete manner: in early 2020, *The Guardian* announced an advertising ban for fossil extractive companies; however, the major weakness of the so-called "fossil ad ban" lies in the fact that for financial reasons, *The Guardian* still accepts advertising from the car and cruise ships industries, i.e. industries closely entangled with the fossil industry. This is only one example of economic pressures reducing the ambition of a CSR strategy. Finally, efforts related to a media company's ecological footprint may include emission assessments (ideally in all 3 scopes), and reduction goals as practised by *The Guardian*, which aims to become carbon neutral by 2030 (The Guardian, 2019). Thus, it is central to remember that—within the constraints of macroeconomic dynamics—the news media as economic institutions have some leverage to make decisions that are either aligned with or stand in direct contrast to their perceived communicative responsibility.

However, while the aforementioned good practices can be observed and may contribute to an extended understanding of news media CSR in the climate crisis, the empirical findings suggest that they tend to do so within niches and, at times, in contradictory ways. It is unlikely that—given contextual macroeconomic dynamics fostering profit orientation and competition—a substantial proportion of media companies will voluntarily act in a manner that is adequate to the extent of the climate crisis. The analysis reveals that "a consistent response to the climate crisis in line with climate science does not seem to be (explicitly) understood as a public responsibility by news media practitioners" and "ethical considerations [. . .] become secondary in light of economic pressures faced by media companies as well as political interests pursued by media owners" (Regen, 2021, p. 100). Yet, listed good practice examples can serve to enrich the imagination of what CSR for news media companies in the twenty-first century could potentially mean.

## 4 The Limitations of CSR

The limitations of CSR as the sole response to the climate crisis are reflected in the following aspects: the voluntary nature of commitments and resulting low public accountability, the potential divergence between intents and realizations, the generic nature of documents linked to CSR intents, the appropriation of CSR as a profit-centred rather than ethics-centred endeavour, and thus the related inferiority of social responsibility declarations to contextual economic and political incentives.

Literature has highlighted the negative potential inherent in CSR measures such as codes of conduct to be little else than "declarations of inten[t]" (Besio & Pronzini, 2014, p. 294). This results in situations in which CSR-based tools such as journalistic codes of conduct "have done little or nothing to prevent unethical practices from occurring" (Couldry, 2013, p. 14). The following example serves as an illustration of how journalistic codes "may be [. . .] little more than professionally required placeholders that may be neglected without stark consequences" (Regen, 2021, p. 102). Both in terms of its communicative and ecological responsibility, News Corp (of which The Australian is one publication) failed to realized its CSR intents. Announcing it would "become the world's first carbon-neutral global media company by 2010" (News Corp, 2008), News Corp removed the sustainability reporting section from its annual reports after it had failed to meet its target, and rebranded its sustainability initiative including less ambitious carbon reduction goals, thereby evading the responsibility to declare its CSR failure publicly. Connected, The Australian's *Editorial Code of Conduct* contains a passage that disallows to "knowingly publish inaccurate or misleading information" (The Australian, 2021): it has been breached on several occasions, including the undermining of IPCC reports' validity, the spread of misinformation on implications of the Paris Agreement, and the publication of opinion pieces on $CO_2$ concentration and climate change by authors who hold close affiliations with the mining industry (Regen, 2021). While voices from civil society and science such as the Union of Concerned Scientists have confronted News Corp, and the Australian Press Council investigated the cases, legal responses to News Corp's misleading reporting behaviour remained largely absent. This illustrates how non-binding documents can be adjusted and appropriated without direct consequences, which we argue is an unreliable means of navigating societal challenges.

An additional challenge emerging from CSR-based documents is their tendentially generic nature as the empirical findings show. The analysis of journalistic codes of conduct and interview data with editors and journalists of *The Guardian*, *Hindustan Times*, and *HuffPost* revealed that—despite all of the three media being committed to quality coverage of the climate crisis—their journalistic codes tend not to include instructions on climate crisis reporting. Considering the fact that the topic of the climate crisis varies largely in character from other issues that are regularly covered, i.e. in terms of its extent, complexity, lack of immediacy, and trend rather than event, arguably, it is advisable for news media committed to engaging with the topic meaningfully to formulate guiding principles related to climate crisis coverage as is the case for other topics, such as in the *Editor's Code of Practice* by the Press Complaints Commission, where particularly sensitive issues are elaborated on.

Finally, the case studies have found CSR to have the potential to become both a) appropriated and b) overshadowed by contextual and personal political and economic factors, with the latter concerning the prioritization of financial incentives over ethical concerns, rendering CSR a profit-centred rather than an ethics-centred endeavour. As regards the former, prior research has revealed that there is a positive correlation between a company's ecological CSR efforts and their readiness for voluntary corporate climate change disclosure (Dawkins & Fraas, 2011); hence, efforts tend to be linked to interests to publicly communicate them and thus are indirectly linked to corporate image and profit. Wang (2014) confirms that whether corporations evaluate CSR positively is dependent on connected profit increases. Empirical findings suggest the same: for news media with a particularly well-defined response to the climate crisis, economic incentives seem to play a central role; in the case of *The Guardian*, a readership-centred business model and readership adjustment appear to be core reasons for its high commitment to the cause. In the case of *The Australian*, political interests (reflected by links to conservative, climate sceptic think tanks) tend to overshadow CSR declarations both in terms of coverage and company operations as has been illustrated above. Finally, *Hindustan Times* shines a light on incongruences between media companies' communicative and ecological responsibility: while publishing high-quality climate crisis coverage, *Hindustan Times* has formed financial partnerships with car corporations to meet the corporation's business needs. Consequently, the context of competitive, profit-centred economies, in which news media's decision-making is guided by economic considerations, is likely to undermine many attempts at attending to news organisations' communicative and ecological responsibility in times of the climate crisis.

## 5 Conclusion and Ways Forward

To conclude, in recent times efforts related to CSR have been made that aim to align news media's actions with challenges posed by the climate crisis; these have been briefly introduced and can enrich the conceptualization of CSR in the news media industry in times of the multiple crises. Examples like *The Guardian* shine light on pioneering CSR activities that reflect new ways of addressing climate breakdown in a corporate setting, and understanding the task both as a communicative and as an ecological responsibility that is connected to a news organization's economic choices. However, CSR alone is insufficient to address the social-ecological challenges presented to contemporary societies as the concept is inextricably intertwined with the logics that have resulted in social and ecological destruction. In fact, CSR does not challenge the existing economic and social relations to a large extent, and thus contributes to their legitimization. Consequently, focussing on CSR in research bears the danger of presenting a distraction from furthering research on media regulation, and cross-sectoral regulation required for social-ecological transformation.

CSR, thus, can at best represent an addition to regulatory efforts required for systemic change that current crises call for; however, it cannot and from an ethical perspective *must* not be viewed as a legitimate substitute to regulation as economic pressures will outweigh moral reasoning in a system that is in itself heavily amoral. System change, e.g., in the form of a Just Transition or a social-ecological transformation reaches beyond what CSR can provide. In contrast, CSR carries the potential to bring about incremental changes, creating the impression of change from within, and thus slowing down regulatory efforts. However, "[w]ithout regulation, responses to the climate crisis will not only likely remain inconsistent and thus construct a contradictory social reality of the climate crisis in the public sphere, but efforts such as those made by The Guardian will also likely remain an exception rather than translate into the standard" (Regen, 2021, p. 107). Thus, we conclude that the call for more and more refined CSR without media regulation will be insufficient in times of intersecting social-ecological crises that require an overarching regulatory framework. Regulation may target media markets, in particular advertising, but also the funding structure for quality news media. Further, when discussing regulations related to the media industry, the latter ought to be viewed as embedded in the structures and dynamics of a growth-based economic system (Knoche, 2014; Mosco, 2009), and reregulation in regard to advertising may in fact be rather realized as a cross-sectoral project. Here it is central that "[t]he complexity of assigning responsibility should not lead to the avoidance of the entanglement of cultural and environmental questions in [media] governance" (Kääpä, 2020, p. 145).

We, further, appeal to a conceptualization of quality news media as a public good, and as a central social-reality shaping component in the public sphere that interlinks otherwise disconnected areas of life. The news media should, thus, not be exposed to the fluctuations of the economy in order to open up long-term spaces for the pursuit of decisions in line with news media's ecological responsibility. Public financing with maximum editorial autonomy and political distance is only one possible way forward. Mixed financing regimes could present an alternative pathway to be further explored in research (c.f. e.g. Meier, 2012). Similarly, the potential of alternative media ownership models should be further investigated as started by Schneider (2021). In addition, the potential of regulation related to transparency requires further research, including "pathways for the disclosure of relations to fossil (based) industries through ownership, advertising, and funding as well as through authors' affiliations in opinion pieces but also in climate science coverage" (Regen, 2021, p. 108). In addition—less so perhaps in the news media industry, but in the entertainment media industry—it would be relevant to look not only into media's "brainprint" (i.e. media content) but also into their "footprint" (Kääpä, 2020).

In a democratic dialogue, the question should be explored "whether the media institutions we have are the media institutions we need, and, if not, what we might, in the interest of living better together, start doing about it" (Couldry, 2013, p. 26). In fact, if quality news media come to be understood as a public good, relevant regulatory efforts ought to be undertaken to free them from the interlinkage with an inherently unsustainable economic system and related fluctuations, notably from the dependence on the advertising industry. This has been a first appeal to turn towards the macro level, and think beyond CSR while appreciating the merits of recent developments. If twenty-first-century organizations are supposed to be "protect[ed] from engaging in illicit behaviour and practices [. . .] that have irrevocable consequences" (Bishop, 2013, p. 637), it would be wise not to leave the decision-making *just* to themselves.

## References

Asara, V., Otero, I., Demaria, F., & Corbera, E. (2015). Socially sustainable degrowth as a social–ecological transformation: Repoliticizing sustainability. *Sustainability Science, 10*(3), 375–384. https://doi.org/10.1007/s11625-015-0321-9

Bashtovaya, V. (2014). Csr reporting in the United States and Russia. *Social Responsibility Journal, 10*(1), 68–84. https://doi.org/10.1108/SRJ-11-2012-0150

Batko, R., & Kreft, J. (2017). The superfluousness of CSR in media organisations. *Journal of Organizational Change Management, 30*(1), 91–105. https://doi.org/10.1108/JOCM-09-2015-0172

Baxter, P., & Jack, S. (2010). Qualitative case study methodology: Study design and implementation for novice researchers. *Qualitative Report, 13*(4).

Beattie, G. (2020). Advertising and media capture: The case of climate change. *Journal of Public Economics, 188*, 1–17. https://doi.org/10.1016/j.jpubeco.2020.104219

Besio, C., & Pronzini, A. (2014). Morality, ethics, and values outside and inside organizations: An example of the discourse on climate change. *Journal of Business Ethics, 119*(3), 287–300. https://doi.org/10.1007/s10551-013-1641-2

Bishop, W. H. (2013). The role of ethics in 21st century organizations. *Journal of Business Ethics, 118*(3), 635–637. https://doi.org/10.1007/s10551-013-1618-1

Boykoff, M., & Boykoff, J. (2007). Climate change and journalistic norms: A case-study of US mass-media coverage. *Geoforum, 38*(6), 1190–1204. https://doi.org/10.1016/j.geoforum.2007.01.008

Boykoff, M. T., & Yulsman, T. (2013). Political economy, media, and climate change: Sinews of modern life. *Wiley Interdisciplinary Reviews: Climate Change, 4*(5), 359–371. https://doi.org/10.1002/wcc.233

Bracker, I. (2017). Corporate social responsibility (CSR) und corporate citizenship (CC): Selbstbild und Fremdwahrnehmung in der öffentlichen Kommunikation. In K.-D. Altmeppen, F. Zschaler, H.-M. Zademach, C. Böttigheimer, & M. Müller (Eds.), *Nachhaltigkeit in Umwelt, Wirtschaft und Gesellschaft: Interdisziplinäre Perspektiven* (pp. 257–284). Springer Fachmedien Wiesbaden; Imprint: Springer VS. https://doi.org/10.1007/978-3-658-14439-5_11

Brand, U. (2009). *Die Multiple Krise. Dynamik und Zusammenhang der Krisendimensionen, Anforderungen an politische Institutionen und Chancen progressiver Politik.* Heinrich Böll Stiftung. https://www.researchgate.net/publication/316890297_Die_Multiple_Krise_

Dynamik_und_Zusammenhang_der_Krisendimensionen_Anforderungen_an_politische_Institutionen_und_Chancen_progressiver_Politik

Brand, U. (2016). How to get out of the multiple crisis? Contours of a critical theory of social-ecological transformation. *Environmental Values, 25*(5), 503–525. https://doi.org/10.3197/096327116x14703858759017

Brand, U., & Wissen, M. (2017). *Imperiale Lebensweise: Zur Ausbeutung von Mensch und Natur im globalen Kapitalismus*. oekom verlag.

Brüggemann, M., & Engesser, S. (2014). Between Consensus and Denial. *Science Communication, 36*(4), 399–427. https://doi.org/10.1177/1075547014533662

Carrington, D. (2015, April 1). Guardian Media Group to divest its £800m fund from fossil fuels. *The Guardian*. https://www.theguardian.com/environment/2015/apr/01/guardian-media-group-to-divest-its-800m-fund-from-fossil-fuels

Carroll, A. B. (1991 (1979)). The pyramid of corporate social responsibility: Toward the moral management of organizational stakeholders. *Business Horizons, 34*(4), 39–48. https://doi.org/10.1016/0007-6813(91)90005-G.

Corfee-Morlot, J., Maslin, M., & Burgess, J. (2007). Global warming in the public sphere. *Philosophical Transactions of the Royal Society of London. Series A, Mathematical and Physical Sciences, 365*(1860), 2741–2776.

Couldry, N. (2013). Why media ethics still matters. In S. Ward (Ed.), *Global media ethics: Problems and perspectives* (1st ed., pp. 13–28). Wiley-Blackwell.

Dawkins, C., & Fraas, J. W. (2011). Coming clean: The impact of environmental performance and visibility on corporate climate change disclosure. *Journal of Business Ethics, 100*(2), 303–322. https://doi.org/10.1007/s10551-010-0681-0

Di Muzio, T. (2015). *Carbon capitalism: Energy, social reproduction and world order*. Rowman & Littlefield International.

Freeman, I., & Hasnaoui, A. (2011). The meaning of corporate social responsibility: The vision of four nations. *Journal of Business Ethics, 100*(3), 419–443. https://doi.org/10.1007/s10551-010-0688-6

Fuchs, C. (2017). Die Kritik der Politischen Ökonomie der Medien/Kommunikation: ein hochaktueller Ansatz. *Publizistik, 62*(3), 255–272. https://doi.org/10.1007/s11616-017-0341-9

Gulyás, Á. (2009). Corporate social responsibility in the British media industries — Preliminary findings. *Media, Culture & Society, 31*(4), 657–668.

Haberl, H., Wiedenhofer, D., Virág, D., Kalt, G., Plank, B., Brockway, P., Fishman, T., Hausknost, D., Krausmann, F., Leon-Gruchalski, B., Mayer, A., Pichler, M., Schaffartzik, A., Sousa, T., Streeck, J., & Creutzig, F. (2020). A systematic review of the evidence on decoupling of GDP, resource use and GHG emissions, part II: Synthesizing the insights. *Environmental Research Letters, 15*(6), 65003. https://doi.org/10.1088/1748-9326/ab842a

Ingenhoff, D., & Bachmann, P. (2017). How do Media companies gain legitimacy? An experimental study on the (Ir)relevance of CSR communication. *International Journal of Strategic Communication, 11*(1), 79–94.

Ingenhoff, D., & Kölling, M. A. (2012). Media governance and corporate social responsibility of media organizations: An international comparison. *Business Ethics a European Review, 21*(2), 154–167.

Kääpä, P. (2020). *Environmental management of the media: Policy, industry, practice (First issued in paperback). Routledge studies in environmental communication and media*. Routledge.

Kalinowska-Żeleźnik, A., Kuczamer-Kłopotowska, S., & Lusińska, A. (2017). The reflections of the CSR strategy in the activities of a public medium, as exemplified by Radio Gdańsk. *Journal of Organizational Change Management, 30*(2), 217–231.

Kannengießer, S. (2020). Nachhaltigkeit und das "gute Leben". *Publizistik, 65*(1), 7–20. https://doi.org/10.1007/s11616-019-00536-9

Karmasin, M. (2006). Medienethik—Normen, Werte und Verantwortung. In C. Scholz (Ed.), *Handbuch Medienmanagement* (pp. 279–297). Springer-Verlag.

Karmasin, M., & Apfelthaler, G. (2017). Integrated corporate social responsibility communication: A global and cross-cultural perspective. In S. Diehl, M. Karmasin, B. Mueller, R. Terlutter, & F. Weder (Eds.), *CSR, sustainability, ethics & governance. Handbook of integrated CSR communication* (pp. 237–250). Springer International Publishing. https://doi.org/10.1007/978-3-319-44700-1_14

Karmasin, M., & Bichler, K. (2017). Corporate social responsibility in the media industry: Setting the benchmark or falling behind? In K.-D. Altmeppen, C. A. Hollifield, & J. van Loon (Eds.), *Value-oriented media management* (pp. 135–146). Springer International Publishing. https://doi.org/10.1007/978-3-319-51008-8_11

Karmasin, M., Kraus, D., Kaltenbrunner, A., & Bichler, K. (2014). Media ethics as institutional ethics – The potential of corporate social responsibility. In S. Fengler, T. Eberwein, G. Mazzoleni, & C. Porlezza (Eds.), *Journalists and media accountability* (pp. 227–245). Peter Lang US.

Karmasin, M., & Litschka, M. (2017). Csr as an economic, ethical and communicative concept. In S. Diehl, M. Karmasin, B. Mueller, R. Terlutter, & F. Weder (Eds.), *CSR, sustainability, Ethics & Governance. Handbook of integrated CSR communication* (pp. 37–51). Springer.

Knoche, M. (2001). Kapitalisierung der Medienindustrie aus politökonomischer Perspektive. *M&K - Medien und Kommunikationswissenschaft*, 177–194. https://uni-salzburg.elsevierpure.com/en/publications/kapitalisierung-der-medienindustrie-aus-polit%C3%B6konomischer-perspek

Knoche, M. (2014). Befreiung von kapitalistischen Geschäftsmodellen: Entkapitalisierung von Journalismus und Kommunikationswissenschaft aus Sicht einer Kritik der politischen Ökonomie der Medien. In F. Lobigs & G. von Nordheim (Eds.), *Journalismus ist kein Geschäftsmodell* (pp. 154–168). Nomos.

Koinig, I., Diehl, S., Weder, F., & Karmasin, M. (2019). Accountability and corporate social responsibility in the media industry: A topic of relevance? In T. Eberwein, S. Fengler, & M. Karmasin (Eds.), *Media accountability in the era of post-truth politics: European challenges and perspectives* (pp. 162–179). Routledge.

Malm, A. (2013). The origins of fossil capital: From water to steam in the British cotton industry. *Historical Materialism, 21*(1), 15–68. https://doi.org/10.1163/1569206X-12341279

McKnight, D. (2010). A change in the climate? The journalism of opinion at news corporation. *Journalism, 11*(6), 693–706. https://doi.org/10.1177/1464884910379704

Meier, W. A. (2012). Öffentlich und staatlich finanzierte Medien aus schweizerischer Sicht. In O. Jarren (Ed.), *Medienstrukturen: Vol. 1. Medienwandel oder Medienkrise? Folgen für Medienstrukturen und ihre Erforschung* (1st ed., pp. 127–145). Nomos. https://doi.org/10.5771/9783845236735-127

Moore, J. W. (2017). The Capitalocene, Part I: On the nature and origins of our ecological crisis. *The Journal of Peasant Studies, 44*(3), 594–630. https://doi.org/10.1080/03066150.2016.1235036

Mosco, V. (2009). *The political economy of communication* (2nd ed.). Sage; ProQuest Ebook Central.

Nerone, J. (2015). *The media and public life: A history (1. publ). New directions in media history*. Polity. http://gbv.eblib.com/patron/FullRecord.aspx?p=1998233

News Corp. (2008). *Annual report 2008*. https://www.annualreports.com/HostedData/AnnualReportArchive/n/NYSE_NWS_2008.pdf

Oreskes, N. (2004). Beyond the ivory tower: The scientific consensus on climate change. *Science (New York), 306*(5702), 1686. https://doi.org/10.1126/science.1103618

Painter, J., & Gavin, N. T. (2016). Climate scepticism in British newspapers, 2007–2011. *Environmental Communication, 10*(4), 432–452. https://doi.org/10.1080/17524032.2014.995193

Paris Agreement. (2015). https://unfccc.int/files/essential_background/convention/application/pdf/english_paris_agreement.pdf

Parrique, T., Barth, J., Briens, F., Kerschner, C., Alejo, K.-P., Kuokkanen, A., & Spangenberg, J. H. (2019). *Decoupling debunked: Evidence and arguments against green growth as a sole strategy*

*for sustainability*. https://www.dnr.de/fileadmin/publikationen/themenhefte/entkopplungsreport_eeb_07_2019.pdf
Ragas, M. W., & Roberts, M. S. (2009). Communicating corporate social responsibility and Brand sincerity: A case study of Chipotle Mexican Grill's 'food with integrity' program. *Journal of Strategic Communication, 3–4*, 264–280.
Regen, L. (2021). *Between carbon capitalism, ethics and science: coverage and operational responses to the climate crisis by selected English-language news media* [Master thesis]. Vienna University of Economics and Business, Vienna, Austria.
Ripple, W. J., Wolf, C., Newsome, T. M., Barnard, P., & Moomaw, W. R. (2019). World scientists' warning of a climate emergency. *Bioscience, 70*(1), 8–12. https://doi.org/10.1093/biosci/biz088
Robin, D. P. (2016). Why ethics and profits can and must work together in business. In R. A. Peterson & O. C. Ferrell (Eds.), *Business ethics: New challenges for business schools and corporate leaders* (pp. 208–233). Routledge. https://doi.org/10.4324/9781315497815-18
Schmidt, A., Ivanova, A., & Schäfer, M. S. (2013). Media attention for climate change around the world: A comparative analysis of newspaper coverage in 27 countries. *Global Environmental Change, 23*(5), 1233–1248. https://doi.org/10.1016/j.gloenvcha.2013.07.020
Schneider, N. (2021). Broad-based stakeholder ownership in journalism: Co-ops, ESOPs, blockchains. *Media Industries Journal, 7*(2). https://doi.org/10.3998/mij.15031809.0007.203
Schwartz, M., & Carroll, A. B. (2003). Corporate social responsibility: A three-domain approach. *Business Ethics Quarterly, 13*(4), 503–530. https://doi.org/10.2307/3857969
Stäbler, S., & Fischer, M. (2020). When does corporate social irresponsibility become news? Evidence from more than 1,000 Brand transgressions across five countries. *Journal of Marketing, 84*(3), 46–67. https://doi.org/10.1177/0022242920911907
Steffen, W., Broadgate, W., Deutsch, L., Gaffney, O., & Ludwig, C. (2015). The trajectory of the Anthropocene: The great acceleration. *The Anthropocene Review, 2*(1), 81–98. https://doi.org/10.1177/2053019614564785
Tench, R., Bowd, R., & Jones, B. (2012). Perceptions and perspectives: Corporate social responsibility and the media. *Journal of Communication Management, 11*(4), 348–370.
The Australian. (2021, March 8). *Editorial code of conduct*. https://www.theaustralian.com.au/editorial-code-of-conduct
The Guardian. (2019). *The Guardian's climate pledge 2019*. https://www.theguardian.com/environment/ng-interactive/2019/oct/16/the-guardians-climate-pledge-2019
The Guardian. (2020, November 20). *Keep it in the ground. The Guardian*. https://www.theguardian.com/environment/series/keep-it-in-the-ground
The Guardian Media Group. (2020). *Annual report and consolidated financial statements for the year ended 29 March 2020*.
Toker, H. (2013). Media CSR Forum. In S. O. Idowu, N. Capaldi, L. Zu, & A. D. Gupta (Eds.), *Encyclopedia of corporate social responsibility* (pp. 1643–1650). Springer.
Trommershausen, A., & Karmasin, M. (2020). Corporate social responsibility in Medienunternehmen. In J. Krone & T. Pellegrini (Eds.), *Handbuch Medienökonomie* (pp. 1235–1296). Springer.
Wang, S. (2014). On the relationship between CSR and profit. *Journal of International Business Ethics, 7/1*, 51–57. http://tarjomefa.com/wp-content/uploads/2017/07/6964-english-tarjomefa.pdf
Yin, R. K. (2018). *Case study research and applications: Design and methods* (6th ed.). Sage.

# Corporate Social Responsibility in German Media Companies: Motivation and Integration into the Corporate Strategy

Annkathrin Clemens and Josephine Franz

## 1 Problem Statement

Today, due to changing societal circumstances (e.g., the loss of traditional social ties (e.g., churches and associations), globalization, or growing privatization), people increasingly turn to companies to identify with (Bennett, 2014; Giddens, 1996; Trommershausen, 2011, p. 87 ff.). As a result, they are gaining further importance as societal institutions, accompanied by a change in their responsibility. Companies must face this challenge for legitimacy, ethical reasoning, and strategic opportunities (Trommershausen & Karmasin, 2016, p. 8). Related corporate strategies and measures can be collected under the term *Corporate Social Responsibility* (CSR). Following a narrow understanding of CSR along the Triple Bottom Line by Elkington (1997) companies have an economic, ecological, and social responsibility. Regarding media companies, they are facing special responsibilities. Their products are not only economic goods, but—in Germany by law—cultural goods and have a special value for society (*public value*). Media companies are thus ascribed a so-called *twin responsibility*: For one, having a responsibility to provide the necessary resources for their editorial content. On the other hand, there is a social, ecological and economic responsibility arising from their corporate activities (Altmeppen, 2011, p. 248; Bracker, 2017, p. 266). Therefore, this study focuses on the media companies' understanding of their responsibility and their strategic approach. With the help of interviews with CSR representatives of media companies, this study aims to determine whether media companies are aware of their twin responsibilities. Additionally, how and why is CSR strategically integrated by German media companies? What are the main goals and motivations for integrating

A. Clemens (✉)
Frankfurt am Main, Germany

J. Franz
Darmstadt, Germany

 
F. Weder et al. (eds.), *CSR Communication in the Media*, CSR, Sustainability, Ethics & Governance, https://doi.org/10.1007/978-3-031-18976-0_10

CSR into corporate strategy and implementing CSR measures? What measures are they already taking? Lastly, is a categorization of the CSR strategy of media companies possible? Can fundamental commonalities be identified in the implementation of CSR in German media companies?

# 2 The Relevance of Responsibility for Media Companies

## *2.1 Responsibility in Media Companies and Their Public Value*

In the public awareness in Germany, the term *media* stands for mass media like newspapers, magazines, radio, and television (Altmeppen, 2006, p. 135). The dual broadcasting system in Germany distinguishes between public and private media companies. Public media companies are financed primarily by broadcasting fees paid by citizens, while private media companies are financed according to the commercial principle, such as advertising revenues. Both are bound by the Broadcasting Treaty though not under same conditions (Die Medienanstalten, 2019). In addition to broadcasting, newspapers and magazines contribute to the media landscape. These are not regulated by state and finance themselves by selling their products to readers and advertising market (Heinrich, 1994, p. 53). Still, they are part of the public understanding of media.

### 2.1.1 Distinction Between Media and Journalism and Their Responsibilities

Based on the public understanding of media described above, media companies are often equated with journalism, but both should be regarded as separate systems. Thus, media companies operate on two levels in society since they produce economic and cultural goods at the same time. Consequently, media companies are divided into two organizational systems, *journalism* and *media*, with different services, orientations, and structures, resulting in other responsibilities for either system (Altmeppen, 2011, p. 251).

The purpose of journalism is to produce current information (Altmeppen, 2006, p. 17) and create publicity so citizens can participate and make decisions in societal processes (Blöbaum, 2000, p. 170). In contrast, the function of media organizations is to bring information offerings (produced by journalism) and entertainment products to the public. To achieve this, capital must be raised, which can subsequently be made available as a resource. Media organizations, therefore, aim to distribute media offerings to enable the business process through financing, procurement, production, and distribution of media (Altmeppen, 2006, p. 17). The activities of media companies thus primarily serve an economic purpose, which is why they are primarily assigned to the subsystem economy (Altmeppen, 2006, p. 143; Kiefer, 2005, p. 21).

Karmasin (2001) looks at media companies from the perspective of communication science and economics and comes to similar conclusions: "Media are a business," but they "produce [...] not only economic goods, but also cultural goods, they produce not only private but also public and quasi-public goods" (Karmasin, 2001, p. 13). He explains further that in the business dimension, media companies are producers and organizations with economic goals. From the communicative dimension, they produce publicity and reflect society. He emphasizes media companies are in a field of tension between economy and communication respectively organization and public (Karmasin, 2001, p. 13). Plus, Sjurts (2005) and Heinrich (2010) point out that media companies are committed to goals of publicity and economy simultaneously (Heinrich, 2010, p. 21 f.; Sjurts, 2005, p. 7).

### 2.1.2 Special Position of Media Companies Through Orientation to the Common Good (Public Value)

Based on the differentiation between journalism and media, social responsibility for journalism is defined as "societal task of [...] producing societally relevant [...] information offerings" (Altmeppen, 2011, p. 257). Here, it is spoken of public value, which is generated to create a public sphere. Public value describes the extent of creating meaning and value for society (Krainer & Weder, 2011, p. 355). In Germany, the Broadcasting Treaty defines that broadcasters are obliged to serve the common good and, thus, generate public value. This legal mandate to create a value contribution to society entrusts them with the political task of informing and creating a public sphere (Die Medienanstalten, 2019). They are "obligated to the free individual and public formation of opinion as well as to the diversity of opinion" (Die Medienanstalten, 2019, p. 6).

### 2.1.3 The Twin Responsibility of Media Companies

The distinction between two organizational systems in media companies results in a *twin responsibility*. On the one hand, the responsibility of media companies lies in financing journalism and provision of resources to ensure a basis for qualitative reporting (Altmeppen, 2011, p. 260). On the other hand, the responsibility arises as an economically operating company (Altmeppen, 2011, p. 260 f.; Ingenhoff & Koelling, 2012, p. 156).

Accordingly, media companies cannot use journalistic products and the public value generated to assume social responsibility for themselves (Altmeppen, 2011, p. 260 f.), even though the responsibility of media companies is often linked to their journalistic products and both levels are undisputedly connected (Gulyás, 2009, p. 659; Toennesen et al., 2013). Altmeppen (2011) points out the differentiation between expectation and commitment. The distribution of journalistic content is societally expected from media and cannot be evaluated as commitment and contribution to CSR (Altmeppen, 2011, p. 261). The responsibility for qualitative reporting is the core of the business and the basic mandate of media (Altmeppen, 2011, p. 264).

Trommershausen (2011) states the legitimacy of media companies "is no longer given by law [. . .] but [. . .] is increasingly provided by public contribution in form of voluntary assumption of social responsibility or increasingly demanded by society" (p. 20). These explanations illustrate that in media companies, a distinction must be made between the societal (journalistic) mandate and the societal (corporate) responsibility in the sense of CSR.

## 2.2 Current Strategic Embedding of CSR in German Media Companies

German media companies have not addressed their corporate responsibility for a long time and have not integrated it into their corporate strategy. In other European media companies, such as the BBC, corporate responsibility and sustainability have been part of the understanding of responsibility for a longer time. In recent years, it can be observed CSR has entered the German media industry (Crome, 2011, p. 259). For a long time, media companies (both private and public) did not attach any value to the issue of CSR. They referred to their public mandate, their public value, and legal regulation in a dual broadcasting system (Winter & Trommershausen, 2011, p. 342). In addition, they often state their commitment to journalistic quality as a societal responsibility and insist on it because "reporting is an essential pillar of their societal legitimacy" (Altmeppen, 2011, p. 255). Several authors consider this view of media companies as not sufficient. They call on media companies to become more aware of their societal responsibility and to address and institutionalize CSR strategically in corporate processes (Altmeppen, 2011, p. 258; Trommershausen & Karmasin, 2016, p. 11).

Studies in recent years have shown the debate on corporate responsibility has intensified in the media industry and different media companies attach varying degrees of importance to CSR. The studies mainly examined the public CSR communication of media companies, i.e., their self-image, by analyzing the content of online information on websites or business reports (Bracker, 2017, p. 270 f.). Studies by von Holly and Stark (2006) and Scholz et al. (2009) showed that CSR did not yet play an important strategic role in German media companies in the early 2000s. However, the media companies were increasingly concerned about their image and were involved in social, ecological, and cultural activities with different focal points. Yet existing potential is only being exploited by a few larger media companies in rudimentary form, e.g., through sustainability reports (Scholz et al., 2009, p. 28 f.). Fully developed CSR strategies could not be identified at this stage

(Scholz et al., 2009; von Holly & Stark, 2006). Even though, the need to integrate CSR conceptually has been recognized (von Holly & Stark, 2006, p. 55ff.).

The results of Bracker (2017) and Bachmann (2017) show that German media companies are addressing their corporate responsibility in recent years but are implementing and anchoring it in their strategy with varying degrees of intensity. Bracker's (2017) results show there are in part strong deviations in the understanding of CSR in individual German media companies and communication strategies pursue different approaches. CSR is close to management in all media companies and responsibilities are not distributed among single employees (Bracker, 2017, p. 274ff.). In his study, Bachmann (2017) classifies media companies according to three legitimation strategies: *Media Responsibility (MR) purists* (focusing on journalistic core activity and associated legitimacy without carrying out specific CSR activities), *MR-/CSR generalists* (attributing MR and CSR to themselves to a high degree and perceived responsibility in journalistic, societal, internal organizational, and ecological terms) and *MR-/CSR dialectics* (perceiving journalistic and economic independence mutually conditioning their success without further voluntary CSR measures) (Bachmann, 2017, p. 200).

## 3 Research Interest and Methodology

Regarding the described current state of research, the aim of this study was to determine the current state of the CSR strategy and activities of media companies in Germany and therefore examine if, why, and how they integrate CSR into their corporate strategy. The focus was on the media companies' understanding of their responsibility and their strategic approach. Are media companies aware of their twin responsibilities? Based on this, what are the motives for integrating CSR into their corporate strategy? In addition, the question of how media companies implement CSR was examined. What measures do they take? Does the CSR strategy at the corporate level also influence the production of content? Regarding the dual broadcasting system in Germany, the study also examined whether the strategy and linking of CSR to the core business differs by looking at private or public media companies. Eventually, the aim was to typologize the CSR strategies of media companies and to arrange them in a uniform scheme. Can fundamental commonalities in the implementation of CSR in German media companies be identified?

For the research project, a qualitative method of partially standardized guideline-based expert interviews was chosen, conducted with CSR executives from media companies. This approach offered the opportunity to obtain detailed information on their CSR activities, associated communication, and overarching strategies behind the CSR activities, as well as on motivation and drivers. Also, insights into their understanding of responsibility and internal organizational structures and resources could be gained in this way.

A catalog of criteria was developed when selecting media companies and contacting their CSR representatives, which formed the basis for the acquisition.

Only German media companies that bundle "in-house and third-party editorial content (informational or entertainment content)" and "carry out direct or indirect distribution" (Wirtz, 2013, p. 16) were interviewed. In addition, due to the dual broadcasting system existing in Germany, private and public media companies were questioned to capture both perspectives. After contacting 40 employees in 28 different media companies, interview partners from seven companies (ARD, ZDF, Hessischer Rundfunk, Mediengruppe RTL, ProSieben/Sat1, Hubert Burda Media, UFA) were finally recruited. The interviews were conducted from June to July 2019.

For the evaluation of the seven interviews, the corresponding transcripts served as raw data for the analysis with the help of qualitative content analysis, according to Mayring (2015).

## 4 CSR in German Media Companies

The interviews with employees of German media companies provided insights into their understanding of responsibility and the motivation behind the implementation of CSR measures in the corporate strategy. In addition, an attempt was made to categorize CSR corporate strategy.

### *4.1 German Media Companies' Understanding of Responsibility*

Media companies have a special responsibility in society because of the content they produce and their activities as an economically active company. The media companies interviewed defined their societal responsibility strongly in terms of public value. Awareness of their *twin responsibilities* as an economic entity was not very widespread.

#### 4.1.1 Public Value: Responsibility Through Produced and Distributed (Journalistic) Content

All media companies were aware of their societal responsibility, which is based on their journalistic products and confirmed a certain expectation placed on them by the public. In this context, all public and private media companies saw their social contribution to democratic cohesion and formation of will and opinion.

Public media companies placed a strong focus on their products and associated public value. In doing so, they emphasized their obligation and fulfillment of the legal mandate "to inform, entertain, advise and educate." Further, they referred primarily to the requirements of the Broadcasting Treaty and distanced themselves from private media companies. Their product is a public value and the "social contribution [...] brand core and raison d'être." Free access to journalistic content and creation of regional and supra-regional content establishes "identity and cohesion in this country"—resulting in their central role for a democratic society and their understanding of responsibility.

Private media companies emphasized their public value by providing information and more specifically by reaching younger audiences. They underlined that private media companies are more likely to reach younger target groups than public media companies and, therefore, this coverage brings a special responsibility. One company stated its commitment goes beyond the legal obligations of the Broadcasting Treaty and its social contribution is seen as going the "extra mile."

Media companies who are publishing and producing entertainment formats emphasized, on the one hand, educating with entertaining content and initiating discussions or contemplations. On the other hand, they offer a public sphere to taboo topics, thus contributing to society while reaching less educated recipients.

Media companies realized a certain expectation from the public toward them and their products. Particularly public media companies perceived a high expectation to "deliver qualitative, attractive and high-quality programming." The fact they are "solidary financed" played a role here, sparking a more vital debate about legitimacy. Private media companies are subject to a code of conduct and felt, due to their reach, an expectation from the public.

### 4.1.2 Twin Responsibility

At corporate level, media companies have a *twin responsibility*, regarding the provision of resources for journalism and having a social responsibility as an economically active company (Altmeppen, 2011, p. 260 f.). Most media companies have recognized a need to assume responsibility at the corporate level to act ecologically, socially, and economically sustainable. However, the interviews gave the impression that many CSR managers were not familiar with the prevailing scientific definition of *twin responsibility*. At least, this has not yet been incorporated into media companies' understanding of responsibility. It was striking how responsibility for society as an economically active company is often equated with journalistic responsibility and journalism services were passed off as CSR services of media. This was already noted by Altmeppen (2011) and Trommershausen and Karmasin (2016) and became apparent during the interviews. For example, a representative of a private media company stated that "when [...] societal responsibility in the sense of disseminating information, promoting democracy, pluralism, diversity of opinion [is mentioned], one has to say that it [...] was day-to-day business without labeling it as CSR." Or a representative of a public media company understood *twin responsibility* of media companies in producing journalistic products and fulfilling the central role as fourth power of the state. Several interview refusals showed such during the acquisition of interview partners, who stated CSR implementation is not necessary for media companies since products fulfill societal responsibility. They are, per se, committed to the common good.

In five out of seven companies, corporate responsibility was primarily related to the content produced and distributed. However, if media companies were explicitly asked about measures, they take for ecological, economic, or social improvements, six out of seven media companies listed specific measures fitting to the classic understanding of CSR apart from journalistic content. One public media company to be highlighted had a clear understanding of its twin responsibility. The interviewee stated, "the topic of CSR, i.e., also consciously [. . .] the triad of social, economic and ecological aspects is also included in [. . .] corporate balance scorecards." This understanding corresponds to the corporate responsibility of the CSR understanding, according to Elkington (1997). The interviewee explained the company's social responsibility includes "handling broadcasting fees sparingly and carefully and using them for the intended purpose of the program." However, it is not only a matter of providing resources for journalism but of "commitments beyond that."

The results show although awareness of corporate responsibility is present in German media companies, its importance is still subordinate. The classic understanding of CSR is not yet prevalent among media companies and they relate the assumption of societal responsibility strongly to their journalistic content.

## *4.2 Motivation and Causes for Implementing CSR in the Corporate Strategy*

The media companies' motivation for implementing CSR in their corporate strategy can be categorized into a societal and corporate levels. The societal level describes a superordinate level addressing the motivation a media company draws from its membership in society. The second level, the entrepreneurial level, summarizes motivations aimed at the company's economic and internal levels.

### 4.2.1 Societal Level

On a societal level, the legitimacy of their company through CSR commitment is a central motivator for almost all media companies. Media companies want to establish credibility and maintain their *license to operate*. For private media companies, this goes hand in hand with a better position in discussions with politicians and associations. Public media companies emphasized their pressure for legitimacy from public funding. Five of seven media companies, with a strong focus on private media companies, want to create a (content-based) value contribution for society and contribute to ecological improvement and development of society. This is because "giving back is our main driver for these CSR projects." Two media companies made it clear that, with their commitment as possible role models, they want to encourage other companies to become more involved. Four media companies cited the zeitgeist and increased stakeholder demands as reasons for their increased commitment. As companies, they are "very strongly driven" by movements such as *Fridays for Future*.

### 4.2.2 Entrepreneurial Level

At the entrepreneurial level, employer branding motivates all interviewees to implement CSR in their corporate strategy and take CSR measures. Media companies registered an increased interest of their employees and applicants in their CSR commitment. Most of the media companies particularly emphasized the area of recruitment. Applicants attach importance to authentic commitment. For media companies, it is a matter of developing an attractive employer brand and differentiating themselves from competitors. Three interviewees emphasized their CSR commitment positively impacts corporate culture and the working environment.

The interviewees differed in their views on the effects of CSR commitment on the competitive situation and profitability. One interviewee from a private media company emphasized it not being a question of economic benefit. Opposing, another named the intensified competitive situation due to internationalization of the media industry. Public media companies mentioned the changing competitive environment, e.g., through streaming services. This has resulted in increased demands of their commitment, which they want to answer with a particular emphasis on public value.

Four private media companies highlighted an intrinsic motivation for being committed to environment and society: "We do all this [...] because it is our conviction." This is due to an understanding of corporate responsibility for sustainable development or personal motivation in management. Many media companies stressed that CSR has often grown at the request of employees. Based on their commitment to the company, CSR has been integrated into the corporate strategy.

### 4.2.3 Causes for the Implementation of CSR

Media companies started their CSR commitment at different times and cited various drivers for implementing CSR in their corporate strategy. One private media company named the initiation of a donation marathon and the creation of a company foundation as the starting point for its CSR commitment. Some misfortunes led another private media company to become involved in foundations. The company's IPO, resulting in reporting obligations and an increased demand deriving from the capital market was cause for a private media company. One private and one public media company stated the issue had grown historically and had no specific reason. A follow-up question from the TV Council led a public media company to begin bundling all CSR activities, strengthening them in terms of personnel and strategy.

Plus, they then published the first CSR report. Only one public media company was unable to name a specific cause.

## *4.3 CSR Strategy in Media Companies*

A key issue in the research process was how CSR is anchored in media companies' corporate values and structures. The aim was to categorize the CSR strategies: Are there generalizations and repeating patterns in the strategic approach of the media companies interviewed? Since some of the media companies were still developing their strategy at the time of the interviews or were not even practicing CSR at all, in most cases, only initial approaches to the integration of CSR could be found.

### 4.3.1 Importance of CSR and Anchoring in Corporate Values

In general, the interviewees' statements on the importance and anchoring of CSR in corporate values coincided with their perception of a *twin responsibility*. CSR plays a vital role in the corporate values of two companies. Two private media companies emphasized its growing importance though it is not yet of the greatest value or has not made much strategic progress. Two public media companies did not see CSR as a core pillar and referred primarily to public value when asked about their societal responsibility. CSR was practiced in passing and sporadically implemented at the time of the interviews but not centrally controlled or organized.

All media companies confirmed the question of whether CSR is anchored in corporate values. Public media companies justified CSR anchoring in their corporate values by legal mandate and social responsibility going hand in hand with it. One public media company additionally emphasized they are committed to social, ecological, and economic responsibilities and implements this through *balance scorecards* with specifically defined targets and measures at all company levels. Private media companies highlighted their commitment to CSR in their corporate values, yet implementation varies. One company has anchored CSR in its corporate values through a code of conduct with guidelines and policies as well as a sustainability strategy including values and principles. Two companies stressed CSR is practiced in the company and is desired by the management. Still, they did not elaborate on how clearly these are formulated: "[...] [W]e simply believe that [...] as a media company it is part of our responsibility to contribute to an intact society and environment [...] [and] that acting responsibly is also not a contradiction [...] to entrepreneurial success." One media company stated that CSR has so far been formulated rather implicitly and is to be anchored concretely in the corporate values in the course of strategic positioning.

### 4.3.2 Integration into the Corporate Structure: Human and Financial Resources

In all media companies interviewed, CSR is located at a higher corporate level and close to management. At private media companies, it is part of corporate communication, while at two public media companies, it is integrated into the corporate organization at administrative level.

Four media companies that are already strategically implementing CSR, or are in the process of doing so, organize themselves in task forces that meet at regular intervals. Members of these task forces are from different departments (Corporate Communications, Compliance, Finance, Legal, Government Relations, and Public Affairs) and almost always have another core occupation. One interviewee described the task forces as coordinator and advisory body consisting of eight employees. Another is named a "Sustainability Board" holding the participation of the Management Board, which defines responsibilities and goals for sustainability issues. In one private media company, the management board is pushing CSR and assigning responsible persons in corporate communications are implementing it. Two public companies have not centralized CSR organization in form of a department or working group yet.

Two media companies (one public and one private) have full-time CSR officers. A CSR officer at the public media company plays a coordination role, is head of the task force and is the company-wide contact person for CSR issues. The CSR officer is supported by a part-time assistant and a consultant from the scientific community. During the research period leading up to the interviews, it became apparent that other media companies have created positions for CSR officers and provided human resources. For example, three publishing houses were identified to have been addressing the issue strategically for a while now and where several employees are responsible for CSR full time. This underlines the steadily growing importance of CSR as a strategic issue, shown in other media companies as well. During the study, those interviewed were responsible for CSR in media companies and used to be based in the fields of communications and political science, law, and business economy. Most have already gained professional experience in corporate communications, public relations, or marketing at (media) companies or agencies. All interviewees have been with their company for some time.

No media company wanted to disclose financial resources used for CSR. One private media company implements many social CSR measures through foundations with "big budgets." Another company emphasized it is willing to distribute content dealing with CSR issues through its channels, even if it is not necessarily popular with consumers: "[T]he general trend is already going in the direction that one (...) [accepts] corresponding losses of market share."

## 4.4 CSR Measures

CSR measures initiated by media companies interviewed can be divided into operative, communicative, and content-related measures and are summarized in Table 1. Four of seven interviewees were able to name measures in all these areas, others in part.

### 4.4.1 Operative Measures

According to Elkington's (1997) narrow understanding of CSR, the operative measures are divided into social, ecological, and economic responsibility. In the area of social commitment, media companies often support regional, social, and cultural projects. In addition, employees launch various initiatives to which they usually have a personal connection or that have arisen from intrinsic motivations. Examples of such initiatives include promoting local cultural institutions, charity runs, the organization of various awards (e.g., to honor social commitment), involvement in refugee aid, or the promotion of children and young people (e.g., through education and aid projects). One media company provides free advertising time for NGOs if advertising space remains free. Social measures mentioned frequently are aimed at employees and their further development. Three companies support voluntary work of their employees and grant free time or honor them with employee awards. Health management is another topic addressed in three companies, e.g., through various sports and fitness offerings or health campaigns. Almost all the companies interviewed mentioned diversity efforts. They promote diversity of employees (age, cultural and religious backgrounds, gender) through networking events for women, equal opportunity officers, or a women's quota. Some companies

**Table 1** CSR Measures of the media companies interviewed (sorted by categories)

| Operative | | | Content | Communicative |
|---|---|---|---|---|
| Social | | Ecological | | |
| • Establishment of foundations<br>• Promotion of social projects<br>• Promotion of young talents<br>• Free advertising for NGOs<br>• Health promotion | • Diversity Programs<br>• HR development<br>• Benefits<br>• Employee engagement | • Green productions<br>• Environmentally friendly use of resources | • Program campaigns<br>• Review of balance in the program by committees | • Stakeholder-communication<br>• Sustainability reports<br>• Reports to the board and committees<br>• Brochures and information material<br>• Campaigns |

cited the promotion of work-life balance (e.g., company daycare centers, parent-child offices, and part-time models), company pension plans, or insurance policies.

The companies interviewed take measures to meet their ecological responsibilities. All of them try to keep their emissions as low as possible. For example, company bicycles or company-owned electric cars, low paper consumption, reusable systems in the canteen, and efficient energy use were cited as specific measures. A public media company that has adopted guidelines for sustainable procurement and monitors its efforts with a $CO_2$ calculator measuring the $CO_2$ footprint (e.g., of productions or company buildings) stood out in particular. This company and a private media company further want to reduce their $CO_2$ emissions with so-called "green productions." The production of media products should be as resource saving as possible, e.g., by avoiding waste, unnecessary printouts, or the efficient use of energy sources. To further promote this project, a so-called "Grüner Drehpass" (Green Production Pass) was to be developed by several organizations (film subsidies, production companies, TV broadcasters, etc.) "which will establish nationwide criteria [. . .] for sustainable production for the first time."

Concrete economic measures to ensure profitability of media companies were not mentioned in the interviews. Nevertheless, streaming and on-demand services, as well as other new online offerings, have changed the market and competitive environment in the media industry in recent years, which was confirmed by some of the companies interviewed: "[. . .] [T]here is no secret [that] [. . .] the media industry [has] changed, due to the large American players that are joining." Accordingly, media companies must adapt to this and expand their portfolio of offerings and develop concepts for the continued existence of the companies. The media companies interviewed described this challenge in part.

### 4.4.2 Communicative Measures

In addition to operational measures, the interviewees also mentioned communicative measures. Two media companies are reporting on their CSR commitment, one regularly since 2016. One public media company published reports on training and further education and was planning a diversity report. In addition, all companies interviewed provided information on responsibility, CSR and ensured transparency on their websites. Internal communication on CSR takes place in varying degrees in all companies interviewed.

When asked whether CSR measures are evaluated, public and two private media companies cited audience editorials, audience discussions, or other external communication channels collecting direct feedback from external stakeholders. Two public media companies named feedback coming through supervisory boards and committees. Three further companies mentioned internal feedback processes, with two private media companies conducting employee surveys or interviews and being accessible to employees through their task forces.

### 4.4.3 Content-Related Measures

In addition to operational and communicative measures, the companies interviewed listed activities that they undertake based on their content-related or journalistic responsibility. According to Altmeppen (2011), media and journalism are to be regarded as two separate systems. Journalistic content is societally expected from media and cannot be counted as a contribution to CSR (Altmeppen, 2011, p. 261). At the same time, media companies and their journalistic products are always associated with each other (Gulyás, 2009, p. 659; Toennesen et al., 2013). Media companies have a special position due to their reach, they influence several million people through their products daily. This goes hand in hand with a special responsibility and corresponding expectations from the public, which all interview partners confirmed. The media companies use this circumstance to pass on responsible content about their products and set agendas. It was clear to all companies interviewed they cannot and do not want to influence journalists, editors, or scriptwriters in their choice of subjects. They have internalized their mission and responsibility for reasons of professional ethics. Nevertheless, some of the CSR managers said they are in contact with editors and are attempting to develop products in collaboration on sustainability topics: "[We are] talking to the main editors, with the editorial heads and are currently considering which pilot projects we can conduct." Three private companies and one public media company stated they draw attention to socially relevant problems with campaigns, theme weeks, and the promotion of individual projects.

## 5 Conclusion

Media companies bear a special *twin responsibility*: As producers of public sphere and distributors of informative content, they make a significant value contribution to society. In addition to their journalistic activities, media companies also act as business enterprises bearing corporate responsibility. The interviews with CSR executives in German media companies showed that not all companies have clearly internalized this awareness of their twin responsibility yet. Most of them primarily referred to their societal responsibility through their editorial content, i.e., public value. However, it became clear awareness of corporate responsibility is growing and many media companies are developing CSR strategies.

Media companies are confronted with growing legitimation pressure and high expectations of their stakeholders and intend to increase their credibility with their CSR measures. A strongly changed competitive environment and changed expectations from the applicant market require media companies to deal with CSR and take action. The representatives emphasized that CSR commitment is very important, especially in the context of employer branding. In addition to changing stakeholder expectations, some intrinsic drivers motivate media companies to engage in CSR:

They want to create value for society and influence sustainable and social improvement with their CSR measures.

Many media companies are currently developing CSR strategies, but others are not addressing them strategically yet. In addition, CSR is given varying degrees of importance within the interviewed companies. Therefore, a categorization of CSR strategy is still difficult. The interviews did not reveal any overarching pattern of strategic approach among the media companies. However, some commonalities and a similar understanding of how media companies approach CSR implementation were identified. For example, several interviewees mentioned task forces in which CSR is centrally managed. In almost all media companies, responsibility for the CSR issue lies with long-term employees and tends to be located at a higher corporate level. In many media companies, more time is already being allocated to the topic or will be in the future. In some cases, dedicated positions have been or will be created for this purpose.

A clear distinction between the CSR efforts of private and public media companies was not possible. It can be said, due to solidarity-based financing, public media companies are constantly under an increased obligation to justify what they spend their money on. Private media companies are freer in the use of funds they generate. It was striking that public media companies tended to define their corporate responsibility in terms of public value. Nevertheless, it must be noted that one of the public media companies has already integrated CSR into its corporate strategy to a considerable extent and is pursuing it professionally. Media companies must, therefore, be viewed independently from the context of private versus public service. Implementation of a CSR strategy was at the time of the interviews at too varying stages in the individual media companies—regardless of whether they are private or public media companies—and is practiced according to different standards.

When asked about measures taken, the media representatives named operational measures, which can be divided into ecological (e.g., green productions) and social commitment (e.g., employee development and diversity efforts). Furthermore, they implement communicative (e.g., brochures and reports) and content-related measures (e.g., exchange with editors).

In summary, some of the media companies interviewed have recognized CSR as an important strategic issue and are developing a strategic approach. At the same time, not all media companies have fully internalized their *twin responsibility*. In some media companies, public value and thus the journalistic value contribution to society was understood as a sufficient assumption of societal responsibility. Therefore, a greater awareness of CSR and its entrepreneurial significance for them still needs to be created. CSR can be approached more strategically and not be seen as *icing on the cake* and hence definitely contribute to economic success (Trommershausen & Karmasin, 2016, p. 4). Nevertheless, the interviews showed many severe efforts by some German media companies to address the topic strategically. Resources have been created for this (e.g., in form of jobs and working hours) and there will be more made available in the future.

# References


Altmeppen, K.-D. (2006). *Journalismus und Medien als Organisationen: Leistungen, Strukturen und Management* (1st ed.). Verlag für Sozialwissenschaften.

Altmeppen, K.-D. (2011). Journalistische Berichterstattung und Media Social Responsibility: Über die doppelte Verantwortung von Medienunternehmen. In J. Raupp, S. Jarolimek, & F. Schultz (Eds.), *Handbuch CSR: kommunikationswissenschaftliche Grundlagen, disziplinäre Zugänge und methodische Herausforderungen; mit Glossar* (1st ed.). VS Verl. f. Sozialwissenschaften.

Bachmann, P. (2017). *Medienunternehmen und der strategische Umgang mit media responsibility und corporate social responsibility*. Springer VS.

Bennett, W. L. (2014). Changing societies, changing media systems: Challenges for communication theory, research and education. In B. Stark, O. Quiring, & N. Jackob (Eds.), *Von der Gutenberg-Galaxis zur Google-Galaxis: Alte und neue Grenzvermessungen nach 50 Jahren DGPuK* (pp. 25–39). UVK Verl.-Ges.

Blöbaum, B. (2000). Organisationen, Programme und Rollen. In M. Löffelholz (Ed.), *Theorien des Journalismus* (pp. 169–183). VS Verlag für Sozialwissenschaften. https://doi.org/10.1007/978-3-322-97091-6_8

Bracker, I. (2017). Corporate social responsibility (CSR) und corporate citizenship (CC): Selbstbild und Fremdwahrnehmung in der öffentlichen Kommunikation. In K.-D. Altmeppen, F. Zschaler, H.-M. Zademach, C. Böttigheimer, & M. Müller (Eds.), *Nachhaltigkeit in Umwelt, Wirtschaft und Gesellschaft* (pp. 257–284). Springer Fachmedien Wiesbaden. https://doi.org/10.1007/978-3-658-14439-5_11

Crome, K. (2011). Nachhaltigkeitsberichterstattung zwischen Programm und Produktion: Über die doppelte Verantwortung von Medienunternehmen. uwf. *UmweltWirtschaftsForum, 19*(3–4), 257–261. https://doi.org/10.1007/s00550-011-0214-z

Die Medienanstalten. (2019). *Staatsvertrag für Rundfunk und Telemedien (Rundfunkstaatsvertrag—RStV). 22.* Fassung. https://www.die-medienanstalten.de/fileadmin/user_upload/Rechtsgrundlagen/Gesetze_Staatsvertraege/Rundfunkstaatsvertrag_RStV.pdf

Elkington, J. (1997). *Cannibals with forks: The triple bottom line of 21st century business.* Capstone.

Giddens, A. (1996). *Konsequenzen der Moderne* (J. Schulte, Trans.). Suhrkamp.

Gulyás, Á. (2009). Corporate social responsibility in the British media industries—Preliminary findings. *Media, Culture & Society, 31*(4), 657–668. https://doi.org/10.1177/0163443709335281

Heinrich, J. (1994). *Medienökonomie: Band 1: Mediensystem, Zeitung, Zeitschrift.* Westdeutscher Verlag.

Heinrich, J. (2010). *Medienökonomie. Band 1: Mediensystem, Zeitung, Zeitschrift, Anzeigenblatt* (3rd ed.). VS Verl. für Sozialwiss.

Ingenhoff, D., & Koelling, A. M. (2012). Media governance and corporate social responsibility of media organizations: An international comparison. *Business Ethics: A European Review, 21*(2), 154–167. https://doi.org/10.1111/j.1467-8608.2011.01646.x

Karmasin, M. (2001). Das Medienunternehmen als kommunikationswissenschaftliches und ökonomisches Betrachtungsobjekt. Konturen einer Theorie der Medienunternehmung. In M. Karmasin, M. Knoche, & C. Winter (Eds.), *Medienwirtschaft und Gesellschaft I. Medienunternehmen und die Kommerzialisierung von Öffentlichkeit* (pp. 11–20). LIT.

Kiefer, M. L. (2005). *Medienökonomik: Einführung in eine ökonomische Theorie der Medien* (2nd ed.). Oldenbourg.

Krainer, L., & Weder, F. (2011). Public Value und die Verantwortung von Medienunternehmen. Corporate Communicative Responsibility als Bedingung für die Möglichkeit der Herstellung von Public Value. In H. Gundlach (Ed.), *Public Value in der Digital- und Internetökonomie* (pp. 355–377). Halem.

Mayring, P. (2015). *Qualitative Inhaltsanalyse: Grundlagen und Techniken* (12th ed.). Beltz Verlag.

Scholz, C., Eisenbeis, U., & Schaaf, N. (2009). Nachhaltigkeit als Ziel und Mittel der Medienunternehmen. uwf. *UmweltWirtschaftsForum, 17*(1), 23–31. https://doi.org/10.1007/s00550-009-0112-9

Sjurts, I. (2005). *Strategien in der Medienbranche Grundlagen und Fallbeispiele* (3rd ed.). Gabler Verlag. http://www.springerlink.com/content/978-3-663-05972-1

Toennesen, C., Hodgson, S., & Mimmack, F. (2013). *Mirrors or movers? Framing the debate about the impact of media content.* https://carnstone.com/downloadInsightFile?insight=3

Trommershausen, A. (2011). *Corporate responsibility in Medienunternehmen.* von Halem.

Trommershausen, A., & Karmasin, M. (2016). Corporate social responsibility in Medienunternehmen. In J. Krone & T. Pellegrini (Eds.), *Handbuch Medienökonomie* (pp. 1–34). Springer Fachmedien Wiesbaden. https://doi.org/10.1007/978-3-658-09632-8_54-1

von Holly, I., & Stark, B. (2006). Gesellschaftliches Engagement deutscher Medienverlage. *CSR-Strategien auf dem Prüfstand. Prmagazin, 36*(12), 61–68.

Winter, C., & Trommershausen, A. (2011). Potenziale und Probleme der Corporate Social Responsibility (CSR) von TIME-Unternehmen unter den Bedingungen der Medienkonvergenz. In H. Gundlach (Ed.), *Public Value in der Digital- und Internetökonomie* (pp. 335–354). Halem.

Wirtz, B. W. (2013). *Medien- und Internetmanagement* (8th ed.). Springer Gabler.

# Public Debates about the Social Responsibility of Media Companies: A Longitudinal Analysis of Swiss Media Companies from 2010 to 2019

**Sarah Marschlich and Daniel Vogler**

## 1 Introduction

Media social responsibility (MSR)—that is, the social responsibility of media companies—has been increasingly discussed, mainly because of their dual role (Adams-Bloom & Cleary, 2009; Olkonen, 2018). Media organizations differ from other organizations insofar as they are committed to the objectives of journalism and the economy while able to produce cultural and economic goods. Until now, the social responsibility of the media has been mainly equated with their journalistic function. Because of increasing competition among media companies and growing moral expectations of organizations and companies in general, the pressure on media companies to assume additional social responsibility has significantly grown (Olkkonen, 2015, 2018).

This chapter refers to MSR as the activities of media organizations through which they take responsibility for society by providing editorial products and cultural goods and engaging in issues of social and media industry relevance, going beyond their value chain and formal legal obligations. We adopt a media studies perspective, considering MSR in the public sphere as negotiated and evaluated by the mass media (see, e.g., Eisenegger & Schranz, 2011; Raupp, 2011). Following the agenda-setting approach, the media can significantly influence the issues and attributions that the public associates with a company (Carroll & McCombs, 2003). Hence, the

S. Marschlich (✉)
Amsterdam School of Communication Research (ASCoR), University of Amsterdam, Amsterdam, The Netherlands
e-mail: s.m.marschlich@uva.nl

D. Vogler
University of Zurich, Forschungszentrum Öffentlichkeit und Gesellschaft (fög), Zürich, Switzerland
e-mail: daniel.vogler@foeg.uzh.ch

 
F. Weder et al. (eds.), *CSR Communication in the Media*, CSR, Sustainability, Ethics & Governance, https://doi.org/10.1007/978-3-031-18976-0_11

expectations of stakeholders and wider society, as reflected in and shaped by mass media communication, influence which responsibilities companies assume to obtain a positive reputation to a great extent (Eisenegger & Schranz, 2011). The salience and evaluation of a company and its social aspects in news media coverage indicate how successful a company is in addressing social expectations, which can positively affect corporate reputation (see Deephouse, 2000; Vogler & Eisenegger, 2020). At the same time, prior research has emphasized that corporate social responsibility (CSR) is often portrayed critically when the company is blamed for some fault (see, e.g., Eisenegger & Schranz, 2011), and this may negatively affect the company's reputation. Although the relationship between CSR communication and reputation has been widely studied (see Ozdora Aksak et al., 2016, for an overview), there is a lack of studies on MSR and its representation and evaluation in the media. Media organizations present a special case regarding their representation in the media because they are both the reporters and reporting subjects.

The current chapter seeks to determine how media companies and their social responsibility are portrayed in the media and to what extent news coverage contributes to the media reputation of these companies. In addition, we investigate whether there are differences between public and private media companies and whether the coverage of MSR varies when media outlets report on their parent media company versus others. This chapter focuses on media companies in Switzerland, providing an interesting case for studying MSR and its link to media reputation. Specifically, the financing of public broadcasting has been the subject of controversial discussions in recent years, including the "No Billag" initiative, which advocated for abolishing the broadcasting fee for the Swiss Radio and Television Corporation (SRG SSR).

To explore MSR coverage and its link to media reputation, we conducted a manual content analysis of media coverage in Swiss media outlets from 2010 to 2019. By providing comprehensive insights into the role of media in negotiating MSR from a long-term perspective, our study extends previous MSR research by analyzing MSR communication at the corporate level (e.g., Hou & Reber, 2011; Ingenhoff & Koelling, 2012; Koinig et al., 2019; Olkkonen, 2015, 2018), significantly contributing to the emerging research field of MSR.

## 2 Literature Review

In this chapter, MSR is viewed from a macro-theoretical media studies perspective; that is, media organizations are seen as embedded in society, and societal demands influence these media organizations' MSR activities and communication to a substantial degree. Media play a crucial role in creating the public sphere, in which organizational behavior is discussed and the perceptions of an organization are shaped (Carroll & McCombs, 2003; Eisenegger et al., 2010). Following this, we argue that the presentation and negotiation of MSR in the media influence the public's perceptions of media organizations. Thus, the evaluation of a corporation in the media, that is, its media reputation (Deephouse, 2000), is a relevant antecedent of stakeholder perception of a corporation (i.e., its corporate reputation) (Eisenegger et al., 2010). Subsequently, we address the specifics of media companies and their social responsibilities. Furthermore, we present conceptual and empirical research on the portrayal of media and CSR in the media and its relation to corporate reputation.

## 2.1 The Role of Media Organizations in Society

Because of technical innovations and the steadily increasing performance of the media, media have penetrated the entire society, influencing its members' perceptions and actions, thus playing a significant role in the social construction of reality (Saxer, 2012). In the media sphere, issues, actors, and organizations are thematized and interpreted, for example, by defining problems and evaluating actions (Entman, 1993) and by attributing (social) responsibilities to certain actors. Following the agenda-setting approach, media play a decisive role in forming the perception and assessment of companies, including their social responsibility (Carroll & McCombs, 2003; Vogler & Eisenegger, 2020). In this way, media organizations become the object of social responsibility attributions.

The responsibility and purpose of media organizations are to produce and distribute journalistic content. In this way, they inherently fulfill an essential social role related to providing information, the formation of public opinion, and public criticism (McQuail, 2003), diversity of output, and the promotion of media literacy (see Gulyas, 2011). Gulyas (2011) noted the universal responsibilities that media organizations should hold, including environmental responsibilities, human rights, community engagement, and labor practices. At the same time, media companies are driven by economic objectives (Karmasin et al., 2014). Indeed, media organizations depend heavily on the advertising market, which is determined by the audience. The more media can acquire readers and viewers, the more efficiently these media organizations can compete in the advertising market. While the nonjournalistic areas of media organizations are influenced primarily by economic logic (e.g., efficiency and competition), editorial departments mostly decide independently of economic logic when it comes to content acquisition, financial funding, and marketing (Altmeppen, 2011; Arnold, 2018). The dualism between financial and journalistic and public responsibility (Adam-Bloom & Cleary, 2009)—or between journalism's ethics and business ethics, as Olkonen (2018) more recently put it—is the distinctive feature of media companies when compared with other organizations. The journalistic and economic roles of media organizations are not necessarily mutually exclusive but are oriented toward each other, and media companies need to find ways to manage their economic and societal impacts by finding a good ethical fit between journalistic ideals and financial goals (Olkonen, 2018; Tsourvakas, 2016).

Because of the transformation of the media landscape, primarily through media convergence and media concentration, as well as the emergence of the Internet and changes in media usage, the economic role of media companies has become more crucial. Therefore, media companies need to organize themselves more efficiently and increase their profits (Karmasin & Bichler, 2017). At the same time, the demands concerning the social responsibilities of companies, including those from the media industry, steadily grow, for instance, when it comes to environmental activities, sustainability, diversity, and other socially relevant issues (Olkonen, 2018). Nowadays, individuals are increasingly concerned about the social impact of companies. Media organizations depend on heterogeneous stakeholders, including customers (especially from the advertising industry), recipients, employees, and citizens. Hence, like other organizations, media organizations face various changing social expectations. These circumstances have led to a change in what MSR comprises (Ingenhoff & Koelling, 2012; Karmasin & Bichler, 2017). In addition to the journalistic function, media organizations assume other social duties (see Koinig et al., 2019; Olkonen, 2018). As Karmasin and Bichler summarized, "[M]edia enterprises are corporate citizens, but to become good corporate citizens, the balance between ethical, democratic, and economic performance must become a core element of their business as well as publishing strategies" (2017, p. 143). Therefore, MSR is increasingly understood as comprising forms of CSR in addition to those issues that are unique for media companies, such as pluralism, media literacy, and responsible advertising (see Olkonen, 2018).

At its core, CSR can be viewed as companies demonstrating that they are going beyond purely profit-making to address a broader set of social, environmental, and ethical concerns (Carroll, 1991; European Commission, 2001). CSR also goes beyond legal compliance and can be considered a company's response to society's increasing expectations of companies, for instance, sustainable production or workplace diversity (Karmasin & Weder, 2011). Previous scholars have applied CSR to media companies (e.g., Ingenhoff & Koelling, 2012; Koinig et al., 2019; Lee et al., 2016; Olkkonen, 2015, 2018; Tsourvakas, 2016). Lee et al. (2016) concluded that media CSR, as they referred to it, comprises CSR and the public interest. Koinig et al. (2019) distinguished between three forms of MSR activity. First, media companies become involved in general CSR topics, for example, community investment, diversity, environmental issues, community investment, or corporate governance. Second, media organizations participate in CSR topics that hold media relevance, including the plurality of opinions, data security, sustainable development of media organizations, compliance, education of citizens, or citizenship. Finally, media organizations engage in media-specific CSR topics, such as the diversity of media coverage, freedom of expression and press, media literacy, transparency, and accountability related to editorial guidelines (Koinig et al., 2019; see also Olkkonen, 2015; Wilenius & Malmelin, 2009). The application of CSR further implies that MSR can occur outside and within the media organization, for instance, in the form of commitment to employees and advocacy for good working conditions (Ingenhoff & Koelling, 2012). Following the reviewed literature, we conceive of MSR as the activities of media organizations through which they take social responsibility by providing editorial products and cultural goods and engage in issues of social and media industry relevance that go beyond their value chain and formal legal obligations.

Empirical research on MSR exceeding the journalistic and democratic role of media organizations has been scant so far, mostly including studies on the prevalence of MSR in corporate communication, including MSR by US media companies (Hou & Reber, 2011), Finnish media organizations (Olkonen, 2018), media organizations across European countries (Ingenhoff & Koelling, 2012; Koinig et al., 2019), and media giants worldwide (Gulyas, 2011). Other scholars have investigated the motives, instruments, and implementation of MSR (Karmasin & Bichler, 2017), external expectations toward the corporate responsibility of media organizations (Olkkonen, 2015), internal perceptions of CSR in media organizations (Gulyas, 2009), and the effect of MSR on the perceptions of media organizations' legitimacy (Bachmann & Ingenhoff, 2017). Finally, critical case studies of media corporations' CSR activities have been conducted (Sandoval, 2014).

Prior research has found that MSR is an increasingly relevant topic for media organizations and that it is communicated through CSR reports (Gulyas, 2011; Hou & Rebler, 2011) and corporate websites (Ingenhoff & Koelling, 2012; Karmasin & Bichler, 2017). MSR topics and the extent of their reporting appear to vary depending on the organization's size, with larger organizations having more extensive CSR reporting (Gulyas, 2009; Hou & Reber, 2011). Moreover, sectoral differences have been found in MSR engagement (Gulyas, 2009). Broadcasting media were found to view MSR much more as an integral part of their business and to engage more in media-specific CSR practices than print media (Gulyas, 2009). However, MSR communication that includes CSR activities is "still not a core issue in Europe's biggest media corporations" (Karmasin & Bichler, 2017, p. 142). As such, 40% of the media organizations in Germany, Switzerland, and Austria still do not communicate any MSR through corporate platforms and channels (Koinig et al., 2019). Others have shown that media organizations often narrow their social responsibility to environmental responsibility (Gulyas, 2009; Olkonen, 2018), failing to "address large areas of social responsibility, whereas 'integrative forerunners' make meaningful connections between journalistic and business responsibilities" (Olkonen, 2018, p. 169). Furthermore, research has indicated that although major media corporations such as Disney and Google have a long-standing CSR tradition and enjoy a high CSR reputation according to international rankings, their actual activities do not correspond to what they communicate or are reputable for (Sandoval, 2014). This inconsistency between CSR communication and CSR practices may cause a severe threat to a company's overall reputation.

Empirical research on the portrayal of MSR in the media is lacking. Studies on the representation of corporate social activities outside the media industry have shown mixed results. On the one hand, studies have indicated that CSR activities are mostly reported in a positive light (Byun & Oh, 2018; Lee & Riffe, 2017). On the other hand, scholars have concluded that media coverage of CSR is predominantly negative (Lee & Carroll, 2011; Vogler & Gisler, 2016). In a long-term study, Lee and Carroll (2011) found that the negative tone in CSR media coverage increased continuously between 1980 and 2004. Likewise, in the case of CSR in Swiss media outlets, a recent study revealed that CSR is covered mostly with a negative valence (Vogler & Eisenegger, 2020). However, how MSR is portrayed in the news media and how the reporting of MSR has developed over time has not yet been sufficiently addressed, leading to the following research question:

*RQ 1: To what extent and how did news coverage about MSR change over time?*

## 2.2 MSR in the Media and Its Contribution to Reputation

How a corporation is evaluated in the news media, i.e., its media reputation (Deephouse, 2000), has been found to have an influence on the evaluation of the corporation by the public (i.e., its corporate reputation) (Carroll & McCombs, 2003; Eisenegger et al., 2010; Vogler & Eisenegger, 2020). Although positive news coverage might enhance reputation, a negative tone in CSR news might decrease corporate reputation (Einwiller et al., 2010). In this chapter, reputation is understood as an organization's status in the public sphere based on the organization's realization of collectively accepted goals and values (Eisenegger, 2015). Reputation comprises different dimensions, including the functional (i.e., competence, performance, and profitability), social (i.e., adherence to norms and values), and expressive (i.e., emotional evaluation of the company) dimensions (Eisenegger, 2015). At the core of social reputation is whether an organization meets social expectations based on social norms and values. Therefore, social reputation encompasses adherence to moral concepts and cultural standards, such as sustainability or the adoption of social responsibility. Thus, MSR can be considered a driver of social reputation and can be expressed through and shaped by media coverage (Eisenegger & Schranz, 2011).

Although MSR and reputation have rarely been examined before, previous research has extensively addressed the relationship between CSR and reputation (for an overview, see Ozdora Aksak et al., 2016). Scholars have found that CSR can secure a company's reputation (Minor & Morgan, 2011) because CSR efforts demonstrate goodwill. Bachmann and Ingenhoff (2017) investigated the relationship between CSR communication and corporate legitimacy, which is closely related to building corporate reputation (Deephouse & Suchman, 2008); they showed that CSR communication (in this case, on a fictitious corporate website) has no impact on corporate legitimacy but positively influences a company's credibility (Bachmann & Ingenhoff, 2017). Mixed results have been found regarding the media coverage of CSR. Although some scholars found that CSR is primarily covered in a positive manner (Byun & Oh, 2018; Lee & Riffe, 2017), other research has indicated predominantly negative CSR coverage (Lee & Carroll, 2011; Vogler & Eisenegger, 2020; Vogler & Gisler, 2016), which may negatively affect corporate reputation (Eisenegger & Schranz, 2011). Therefore, CSR communication can become a reputational risk. However, a recent study found that even though CSR reporting in legacy media was mainly negative, the salience and tone of news media coverage about CSR overall still increased corporate reputation (Vogler & Eisenegger, 2020). In summary, previous studies have shown mixed results on CSR (outside the media industry) and reputation. However, the role that media coverage of the social responsibility of media organizations plays in forming organizational reputation is not yet clear. Therefore, we pose the following research question:

*RQ 2: To what extent and how does news coverage about MSR contribute to media companies' media reputation?*

MSR can be applied to private and public media companies. In comparison, the focus of public service media companies is not primarily on making a profit but on providing a public service because these companies are predominantly financed by public funds. Therefore, the media may discuss the social role of public and private media companies differently. Concerning MSR, previous studies have indicated a considerable variance between private and public media organizations (Ingenhoff & Koelling, 2012). Accordingly, public media organizations communicate MSR more extensively than public organizations, "which may be because of the fact that public service media organizations feel more accountable to the public as they have a public service obligation" (Ingenhoff & Koelling, 2012, p. 154). Second, private media companies in Switzerland were found to be more concerned with traditional CSR topics, such as sustainability at the economic, ecological, and social levels, while public media companies referred more to their educational obligation and media governance issues (Koinig et al., 2019). To investigate whether the news coverage about MSR differs between the MSR efforts of public and private media organizations, we raise the following research question:

*RQ 3: Are there differences in the news coverage about MSR and media reputation between public and private media companies?*

Companies face the dilemma that, on the one hand, they are expected to take on social responsibilities, yet on the other hand, the more they disclose their social commitment, the more they are met with skepticism (Bachmann & Ingenhoff, 2017), which is described as the CSR paradox (Morsing et al., 2008). For this reason, it is a significant advantage for companies if independently appearing actors such as journalists report (positively) on MSR because this reduces the skepticism of the recipients (Morsing et al., 2008). In turn, this increases the chance of building a good reputation and legitimacy (Bachmann & Ingenhoff, 2017; Eisenegger & Schranz, 2011). However, media organizations are unique compared with other organizations because they must fulfill what society expects of public communication in the media (e.g., critical and objective reporting) and position themselves toward society as responsible and committed organizations (Altmeppen, 2011). The latter can be communicated through their own and external media channels. Hence, by employing "media public relations," media organizations can use their journalistic products for a positive self-portrayal and instrumentalize editorial departments to enforce self-interests in terms of organizational public relations (Arnold, 2018). Media, especially legacy media, enjoy a high level of credibility and legitimacy because of their established role in society (McQuail, 2010). For this reason, we assume that the self-presentations of media organizations in the media might be perceived less as serving organizational self-interests compared with other companies. This particularity puts media organizations in a privileged position to enhance their reputation.

Previous research has shown that the presentation and evaluation of the content relating to one's own media organization are significantly less critical (Lichtenstein, 2011). Likewise, Kemner et al. (2008) revealed that the intensity and evaluation of reporting on media organizations are predominantly carried out according to the interests of the media organizations' publishers. To support the publishers' positions, voices external to the organization are used within the reporting in owned media outlets (Kemner et al., 2008). When the media reports on media companies, not only do economic goals play a role, but anything related to the self-interests of the media organizations (Arnold, 2018). In this way, the journalistic role may be used to influence public opinion formation in favor of the objectives of media organizations (Löblich, 2011), turning media into strategic actors (Brüggemann et al., 2016). This phenomenon can be observed particularly in times of crisis, where media use crisis frames to demonstrate their social role, which must be preserved to justify their privilege in society (Brüggemann et al., 2016).

To sum up, media organizations may use their journalistic role to present themselves as relevant to society by positively reporting on the MSR of their own media organization. At the same time, however, media companies may present less about their own MSR efforts because they fear being met with skepticism (see Morsing et al., 2008), which leads to our final research question.

*RQ 4: Are there differences in the news coverage regarding MSR by owned media outlets and media outlets of other media companies?*

## 2.3 Study Context: Media Corporations in Switzerland

Prior research on MSR and Swiss media organizations is scarce, with a few notable exceptions (Ingenhoff & Koelling, 2012; Koinig et al., 2019). Koinig et al. (2019) investigated the relevance and topics of CSR in German, Swiss, and Austrian media. The authors conducted a content analysis of the media organizations' online communication (such as corporate websites and reports), finding that two-thirds of Swiss media companies engage in MSR efforts, predominantly related to engagement, governance, and sustainability (Koinig et al., 2019). At the same time, however, the study showed considerable variance between private and public media companies. Private media companies, including Ringier and Publigroupe/Swisscom, dealt more with traditional CSR topics, such as sustainability, climate change mitigation, and fair value creation. The public media company SRG SSR, on the other hand, referred more to its educational obligation, which was addressed under the term governance as the social responsibility of the company. In addition, SRG SSR referred to economic conditions and its market share in the area of its MSR communication, along with its commitment to the topics of culture, entertainment, and sports (Koinig et al., 2019). Likewise, Ingenhoff and Koelling (2012) examined the MSR activities of media organizations in European countries, including Switzerland, by performing content analyses of corporate-led communication. In line with Koinig et al. (2019),

Ingenhoff and Koelling (2012) found differences between public and private media organizations, implying that public media organizations communicate their MSR efforts considerably more than private media companies.

Switzerland is a small and highly concentrated media market. A few media companies own most of the news media outlets that have a high reach. This includes a handful of private media companies and the public service broadcaster SRG SSR. Over the past few years, as in many countries, media companies in Switzerland have increasingly come under pressure because of shrinking subscription and advertising revenues (fög—Forschungszentrum Öffentlichkeit und Gesellschaft, 2020). The economic pressure on media companies has led to political pressure on the media system and public service broadcasters in particular. Worth mentioning here is the so-called "No Billag" initiative, which in 2018 demanded abolishing the mandatory license fee for broadcasting. Even though a majority rejected the initiative, it represented a crucial incident for journalism in Switzerland that would have fundamentally changed the media system (Udris et al., 2020).

## 3 Method

By conducting a quantitative manual content analysis of news coverage from the four largest Swiss media companies—SRG SSR, TX Group (formerly Tamedia), Ringier, and NZZ Media Group—in 11 Swiss media outlets, we investigated the importance of media companies' social responsibility in news media coverage and how it has developed over time (2010–2019). The media sample included quality newspapers ($n = 2$), tabloids ($n = 2$) and Sunday papers ($n = 3$), weekly magazines ($n = 2$), and the most-watched newscasts from the public service broadcaster SRG SSR ($n = 2$).

We retrieved all news articles containing references to at least one of the media companies. We only included that coverage mentioning the name of the media company (e.g., Ringier); that is, we did not include coverage referring to the media outlets of the analyzed companies (e.g., the tabloid *Blick*, which is owned by Ringier). We subsequently coded all articles covering a media company in at least one section or in more than one-third of an article. In some news articles, more than one company was mentioned. Our final sample consisted of 4851 codings of media companies in 4556 news articles. Print media articles were accessed through the Swiss Media Database (SMD). The TV newscasts were retrieved from SRG SSR's website.

First, we coded whether the article mainly referred to a media company's social responsibility. We used a broad understanding of MSR and included all debates focusing on the aspects of the behavior of media companies with a social impact (compared with functional economic aspects). Thus, the MSR category essentially included articles on the CSR activities of media companies, articles on media policy issues with the media companies involved (e.g., the No Billag referendum), articles on the legal issues of media companies, or articles on ethical issues and social aspects of media companies' business. Second, we coded the evaluation of the media company in the news article by distinguishing between positive, negative, and balanced or neutral evaluations. Trained coders conducted the manual content analysis. Intercoder reliability was tested with a sample of 100 unique articles coded by two coders. Krippendorff's alpha was 0.94 for the variable reference to MSR and 0.82 for the variable tone, indicating high reliability.

Furthermore, we were interested in the differences between coverage of the social responsibility of the public service broadcaster compared with private media organizations. Therefore, we grouped together the three private media companies—TX Group, Ringier, and NZZ Media Group—and compared them with the public service broadcaster SRG SSR. Finally, for every article on a company, we automatically assigned whether it appeared in an owned outlet or non-owned outlet to examine how media outlets cover the MSR of their parent media company. The procedure resulted in a binary variable indicating whether an article appeared in an owned (1) or non-owned outlet (0). For this step, we additionally determined the owner of the 11 analyzed outlets through desktop research.

## 4 Results

In the following sections, we present the results of our empirical investigation of the news coverage of Swiss media companies' MSR.

### *4.1 MSR Gains Importance in the News over Time*

To answer our first research question, we investigated how the share of MSR in news coverage has developed over time. On average, MSR was mentioned in about one out of three articles (30.5%) and was between 20.4% in 2010 and 43.6% in 2018. MSR seems to have gained importance in Swiss news media coverage over time. However, this trend appears to be heavily driven by political debates surrounding public service broadcasters. The news focusing on MSR was higher for public service broadcasters (PSBs) than for private companies, especially in years with political referendums or initiatives related to the media system. The referendum on the revision of the law on radio and television in 2015 and the "No Billag" initiative in 2018 led to increased visibility of MSR in the media. Thus, political initiatives or referendums related to the media system tended to increase the importance of MSR for media companies but only for the PSBs. The shares of MSR coverage for private media companies were the lowest in the years of the "No Billag" initiative.

## 4.2 Evaluation of MSR Becomes Less Negative Over Time

Concerning reputational aspects, we asked how the MSR of media companies was evaluated in the coverage. Overall, we found significantly more negative evaluations in MSR coverage ($M = -24.4$, $SD = 60.8$, $n = 1478$) than in that of functional or economic topics ($M = -5.4$, $SD = 55.1$, $n = 3373$; $t(4849) = 10.701$, $p < 0.001$). Over the years, the evaluation of MSR has become less negative. In the years of the referendum and initiative, the evaluations were less negative than in other years. Over time, our analysis has revealed differences between private media companies and the PSBs, yet without any clear pattern. Private media companies seem to be evaluated slightly more negatively than PSBs, even though they were evaluated more positively in 2015 and 2016. In the year of the "No Billag" initiative, the MSR of the PSBs was covered less negatively than in the other years.

## 4.3 Coverage of MSR More Often and Less Positively in PSBs than in Private Media Companies

For our third research question, we compared the share of MSR reporting (as used for RQ1) and the evaluation of MSR in reporting (as used for RQ2) between private and public media companies. Our analysis showed that MSR played a more important role in reporting on PSBs (40.2%) than on private media companies (16.9%) ($\chi^2 = 302.2$; $df = 1$; $p < 0.001$). Regarding the evaluation, we found that MSR reporting is less negative for PSBs ($M = -21.0$, $SD = 58.0$, $n = 1137$) than for private media companies ($M = -35.5$, $SD = 68.2$, $n = 341$; $t(1476) = -3.870$, $p < 0.001$).

## 4.4 Coverage of MSR Less Frequent and More Positive in Owned Outlets than in Nonowned Outlets

For our fourth research question, we compared the share of MSR reporting (as used for RQ1) and the evaluation of MSR in reporting (as used for RQ2) in owned and non-owned media outlets. Our analysis showed that the MSR of a company played a less important role in the coverage of owned (11.6%) than of non-owned outlets (34.9%) ($\chi^2 = 101.7$; df $= 1$; $p < 0.001$). This means that media outlets more often report about the MSR of other companies than about the MSR of their parent companies. Regarding the evaluation, our analysis revealed that the reporting of a company's MSR is positive in owned outlets ($M = 7.5$, $SD = 71.0$, $n = 107$) but negative in non-owned outlets ($M = -26.8$, $SD = 59.3$, $n = 1371$; $t(1476) = -5.681$, $p < 0.001$). This result implies that media outlets report more negatively or critically about the MSR of other companies than about the MSR of their parent companies.

## 5 Discussion and Conclusion

This chapter aimed to investigate the coverage of MSR by Swiss media organizations in the Swiss media to find out to what extent MSR coverage influences the media reputation of media organizations and what differences can be observed between private and public media organizations, as well as between own media outlets and those of other media companies. First, the study showed that MSR coverage increased from 2010 to 2019, despite a slightly declining trend since 2017. However, the results indicated that MSR coverage of public media organizations accounts for the increase in scope, which may be explained by the critical debate about the public broadcasting fee because of the "No Billag" initiative. Hannah and Zatzick (2008) pointed out that events such as corporate misdeeds can drive reporting on social and moral issues. However, our results only allow for implicit assumptions about the role of public debates on media organizations in MSR. Therefore, further research is needed to investigate the relevance of events and public discussions for MSR coverage.

Second, MSR coverage was predominantly found to be negative in connotation; that is, MSR was viewed mostly critically in the media. This result is in line with previous findings on the media coverage of CSR (e.g., Vogler & Eisenegger, 2020). The negative tone of CSR and MSR media coverage could be explained by the increase in critical and scandalous news about corporations (Barkemeyer et al., 2020). As such, Eisenegger and Schranz (2011) pointed out that reporting on CSR mainly occurs when the social role of media organizations is viewed critically and when organizations are accused of not fulfilling their ascribed social responsibility. Further studies are necessary to better understand the role of the increasing scandalization of business reporting and MSR and its implications for media organizations. It would be of great interest to investigate MSR topics covered in the media to understand whether some MSR efforts or social aspects of media organizations are presented more critically than others.

However, for both public and private media organizations, it became evident that the negative tone of MSR reporting decreased between 2010 and 2019, showing a downward trend since 2016 for private media organizations and in 2019 for public media organizations. These results again point to the role of public debates and events in determining MSR coverage. As such, the rejection of the "No Billag" initiative in 2018—and, in this way, the approval of the role of public media organizations—might have led to a more positive discussion of the social role of public media organizations. Furthermore, the critical negotiation of the role of public media organizations in the wake of the "No Billag" initiative and the preceding coverage may also have led to more positive media coverage of the social role of public media organizations in 2019, explaining this outlier. Overall, our results indicate differences in MSR news coverage between public and private media organizations. Bringing this together with previous research, it seems that there are differences between public and private media organizations in (organization-led) MSR communication by Swiss media organizations (Ingenhoff & Koelling, 2012) and in the reporting of these efforts. However, with our methodology, we cannot draw explicit conclusions about the relationship between organizational MSR communications and MSR reporting. Future research is required to examine the differences between public and private media organizations regarding MSR efforts and media coverage, along with how and why these differences emerge.

Third, our results imply that through its predominantly negative coverage, MSR may pose a risk to the social reputation of media organizations. Primarily, this result indicates that the reputational logics of companies in other industries is applicable to media companies, underlining the role of efficient media management in securing social acceptance—something that holds true for both public and private media organizations (see Sigler, 2010). Critical reporting on the social role of public media organizations entails the risk of diminishing social approval, which is fundamental to public service media organizations (see Wenzel, 2012). These results suggest that, similar to the organizations of other sectors, media organizations should increasingly monitor how they are portrayed in the media to respond effectively. However, it should be noted that an organization's reputation refers to the social aspects *and* functional and expressive aspects (Eisenegger, 2015). Furthermore, we have considered only the constitution of reputation in legacy media. Other media forms, including digital and social media, individual experiences, and interpersonal communication, play an additional role in the constitution of media reputation.

Finally, our results show that MSR is reported significantly less but more positively in owned media outlets than in non-owned ones. This finding is surprising because we assumed that media organizations present themselves as socially responsible in their outlets. However, when considering the CSR paradox (Morsing et al., 2008), they may fear causing skepticism among recipients or other media outlets. However, we should note that our data cannot make explicit assertions about whether the media organizations reported on actual MSR efforts or their general social role. Follow-up research is needed to examine this more closely and, for example, to include the organizations' perspectives to better understand what opportunities and challenges media organizations associate with MSR.

Because of the broad conceptualization of MSR as the activities of media organizations that have a social impact, our results are limited. Any media reference considering media organizations on a social level (as opposed to functional economic aspects) was collected. Consequently, we cannot conclude how MSR measures initiated by media organizations are discussed and evaluated in the media. Instead, our study shows how the social role of media organizations is debated in media outlets. For subsequent studies, it would be relevant to analyze explicit references to those MSR measures initiated by media organizations and reported in the media and to compare them with our findings. Furthermore, a limitation of the current study is the assessment of reputation based on media reports. There is consensus in the literature regarding the role of the media in the constitution of corporate reputation (Carroll & McCombs, 2003; Deephouse, 2000; Eisenegger & Schranz, 2011). However, it is also evident that other factors besides media coverage can influence the perceptions of companies, primarily through the digital transformation of organizational communication. Nevertheless, recent research shows that legacy media can influence corporate reputation more than corporate communication on social media (Vogler & Eisenegger, 2020), which is also related to the media's perception as independent and credible actors. Finally, the specific country context is a limitation. In the case of Switzerland, diverse and partly critical debates on the social role of (public) media organizations have occurred during the period under study, which certainly affected the media coverage of MSR. Moreover, it can be assumed that the media system and the specific (journalistic) culture influence how MSR is portrayed in the media. For this reason, further research on the coverage of MSR in other country contexts is needed to gain insights into the role of contextual factors in media representations and evaluations of MSR.

By providing valuable and comprehensive insights into the role of the media in negotiating MSR, our study significantly contributes to the emerging field of MSR research. In adopting a macro-theoretical approach to MSR, our study extends previous empirical MSR studies that have focused on MSR communication at the corporate level (Gulyas, 2009, 2011; Ingenhoff & Koelling, 2012; Koinig et al., 2019; Olkkonen, 2015, 2018), thereby responding to the call for media studies on MSR (Ingenhoff & Koelling, 2012). The relevance of our study lies predominantly in its longitudinal research design, which allows for analyzing MSR media coverage in its development and reflecting on the role of public debates and events for MSR. Furthermore, we contribute to CSR research by shedding light on public discussions of the social responsibility of organizations of a specific industry in the media.

## References

Adams-Bloom, T., & Cleary, J. (2009). Staking a claim for social responsibility: An argument for the dual responsibility model. *International Journal on Media Management, 11*(1), 1–8.

Altmeppen, K.-D. (2011). Journalistische Berichterstattung und Media Social Responsibility: Über die doppelte Verantwortung von Medienunternehmen. In J. Raupp, S. Jarolimek, & F. Schultz (Eds.), *Handbuch CSR. Kommunikationswissenschaftliche Grundlagen, disziplinäre Zugänge und methodische Herausforderungen* (pp. 247–266). Springer.

Arnold, D. (2018). Das Vertrauensdilemma im Verhältnis zwischen Medien-PR und Medienjournalismus. In O. Hoffjann & R. Seidenglanz (Eds.), *Allmächtige PR, ohnmächtige PR* (pp. 175–196). Springer.

Bachmann, P., & Ingenhoff, D. (2017). How do media companies gain legitimacy? An experimental study on the (Ir)relevance of CSR communication. *International Journal of Strategic Communication, 11*, 79–94.

Barkemeyer, R., Faugère, C., Gergaud, O., & Preuss, L. (2020). Media attention to large-scale corporate scandals: Hype and boredom in the age of social media. *Journal of Business Research, 109*, 385–398.

Brüggemann, M., Humprecht, E., Kleis Nielsen, R., Karppinen, K., Cornia, A., & Esser, F. (2016). Framing the newspaper crisis. *Journalism Studies, 17*, 533–551.

Byun, S. K., & Oh, J. M. (2018). Local corporate social responsibility, media coverage, and shareholder value. *Journal of Banking & Finance, 87*, 68–86.

Carroll, A. B. (1991). The pyramid of corporate social responsibility: Toward the moral management of organizational stakeholders. *Business Horizons, 34*, 39–48.

Carroll, C. E., & McCombs, M. (2003). Agenda-setting effects of business news on the public's images and opinions about major corporations. *Corporate Reputation Review, 6*, 36–46.

Deephouse, D. L. (2000). Media reputation as a strategic resource: An integration of mass communication and resource-based theories. *Journal of Management, 26*, 1091–1112.

Deephouse, D. L., & Suchman, M. (2008). Legitimacy in organizational institutionalism. In R. Greenwood, C. Oliver, R. Suddaby, & K. Sahlin (Eds.), *The SAGE handbook of organizational institutionalism* (pp. 49–77). SAGE.

Einwiller, S. A., Carroll, C. E., & Korn, K. (2010). Under what conditions do the news media influence corporate reputation? The roles of media dependency and need for orientation. *Corporate Reputation Review, 12*, 299–315.

Eisenegger, M. (2015). Identität, Image und Reputation – Eine kommunikationssoziologische Begriffsarchitektur. In R. Fröhlich, P. Szyszka, & G. Bentele (Eds.), *Handbuch der Public Relations. Wissenschaftliche Grundlagen und berufliches Handeln* (pp. 431–460). Springer.

Eisenegger, M., & Schranz, M. (2011). CSR – Moralisierung des Reputationsmanagements. In J. Raupp, S. Jarolimek, & F. Schultz (Eds.), *Handbuch CSR. Kommunikationswissenschaftliche Grundlagen, disziplinäre Zugänge und methodische Herausforderungen* (pp. 71–96). Springer.

Eisenegger, M., Schranz, M., & Schneider, J. (2010). Corporate reputation and the news media in Switzerland. In C. E. Craig (Ed.), *Corporate reputation and global news media: Case studies of agenda-setting* (pp. 207–220). Routledge.

Entman, R. M. (1993). Framing: Toward clarification of a fractured paradigm. *Journal of Communication, 43*(4), 51–58.

European Commission (2001). *GREEN PAPER promoting a European framework for corporate social responsibility*. Available from europa.eu/rapid/press-release_DOC-01-9_en.pdf.

Fög – Forschungszentrum Öffentlichkeit und Gesellschaft. (2020). *Jahrbuch 2020 Qualität der Medien. Schweiz – Suisse – Svizzera*. Schwabe.

Gulyas, A. (2009). Corporate social responsibility in the British media industries - preliminary findings. *Media, Culture & Society, 31*, 657–668.

Gulyas, A. (2011). Demons into angels? Corporate social responsibility and media organisations. *Critical Survey, 23*, 56–74.

Hannah, D. R., & Zatzick, C. D. (2008). An examination of leader portrayals in the US business press following the landmark scandals of the early 21st century. *Journal of Business Ethics, 79*, 361–377.

Hou, J., & Reber, B. H. (2011). Dimensions of disclosures: Corporate social responsibility (CSR) reporting by media companies. *Public Relations Review, 37*(2), 166–168.

Ingenhoff, D., & Koelling, A. M. (2012). Media governance and corporate social responsibility of media organizations: An international comparison. *Business Ethics: A European Review, 21*, 154–167.

Karmasin, M., & Bichler, K. (2017). Corporate social responsibility in the media industry: Setting the benchmark or falling behind? In K.-D. Altmeppen, C. A. Hollifield, & J. van Loon (Eds.), *Value-oriented media management* (pp. 135–146). Springer International Publishing.

Karmasin, M., Diehl, S., & Koinig, I. (2014). Medienökonomie und Medienmanagement als Integrationsdisziplinen – Vom Objektbereich zur Heuristik ? In M. Karmasin, M. Rath, & B. Thomaß (Eds.), *Kommunikationswissenschaft als Integrationsdisziplin* (pp. 75–93). Springer Fachmedien.

Karmasin, M., & Weder, F. (2011). CSR nachgefragt: Kann man Ethik messen? In J. Raupp, S. Jarolimek, & F. Schultz (Eds.), *Handbuch CSR* (pp. 463–479). VS Verlag für Sozialwissenschaften.

Kemner, B., Scherer, H., & Weinacht, S. (2008). Unter der Tarnkappe. *Publizistik, 53*, 65–84.

Koinig, I., Voci, D., Weder, F., & Karmasin, M. (2019). Media social responsibility an der Schnittstelle von Media accountability und corporate social responsibility: Ein theoretisches Konzept und eine empirische Untersuchung der Medienindustrie am Beispiel der D-A-CH Länder. In M. Litschka & L. Krainer (Eds.), *Der Mensch im digitalen Zeitalter* (pp. 103–133). Springer Fachmedien.

Lee, S. Y., & Carroll, C. E. (2011). The emergence, variation, and evolution of corporate social responsibility in the public sphere, 1980–2004: The exposure of firms to public debate. *Journal of Business Ethics, 104*, 115–131.

Lee, T. H., & Riffe, D. (2017). Business news framing of corporate social responsibility in the United States and the United Kingdom: Insights from the implicit and explicit CSR framework. *Business & Society, 58*, 683–711.

Lee, C. G., Sung, J., Kim, J. K., Jung, I. S., & Kim, K. J. (2016). Corporate social responsibility of the media: Instrument development and validation. *Information Development, 32*, 554–565.

Lichtenstein, D. (2011). Kommerzialisierung des Medienjournalismus? Eine empirische Untersuchung zum "Fall Berliner Zeitung". *Medien & Kommunikationswissenschaft, 59*, 216–234.

Löblich, M. (2011). Frames in der medienpolitischen Öffentlichkeit: Die Presseberichterstattung über den 12. Rundfunkänderungsstaatsvertrag. *Publizistik, 56*, 423–439.

McQuail, D. (2003). *Media accountability and freedom of publication*. Oxford University Press.

McQuail, D. (2010). *McQuail's mass communication theory*. SAGE.

Minor, D., & Morgan, J. (2011). CSR as reputation insurance: Primum non Nocere. *California Management Review, 53*, 40–59.

Morsing, M., Schultz, M., & Nielsen, K. U. (2008). The 'catch 22' of communicating CSR: Findings from a Danish study. *Journal of Marketing Communications, 14*, 97–111.

Olkkonen, L. (2015). Audience enabling as corporate responsibility for media organizations. *Journal of Media Ethics, 30*, 268–288.

Olkonen, L. (2018). Separation or integration of journalistic and business responsibilities? Corporate responsibility in Finnish media companies. *Journal of Media Business Studies, 3*, 169–187.

Ozdora Aksak, E., Ferguson, M. A., & Atakan Duman, S. (2016). Corporate social responsibility and CSR fit as predictors of corporate reputation: A global perspective. *Public Relations Review, 42*, 79–81.

Raupp, J. (2011). Die Legitimation von Unternehmen in öffentlichen Diskursen. In J. Raupp, S. Jarolimek, & F. Schultz (Eds.), *Handbuch CSR* (p. 97). VS Verlag für Sozialwissenschaften.

Sandoval, M. (2014). *From corporate to social media. Critical perspectives on corporate social responsibility in media and communication industries*. Routledge.

Saxer, U. (2012). *Mediengesellschaft. Eine kommunikationssoziologische Perspektive*. VS Verlag für Sozialwissenschaften.

Sigler, C. 2010. Online-Medienmanagement. Grundlagen – Konzepte – Herausforderungen. Mit Praxisbeispielen und Fallstudien. : Gabler.

Tsourvakas, G. (2016). Corporate social responsibility and media management: A necessary Symbiosis. In G. F. Lowe & C. Browns (Eds.), *Managing media firms and industries* (pp. 143–158). Springer.

Udris, L., Eisenegger, M., Vogler, D., Schwaiger, L., & Häuptli, A. (2020). Reporting the current media system is at stake: Explaining news coverage about the initiative on the abolition of public service broadcasting in Switzerland. In E. C. Tandoc, J. Jenkins, R. J. Thomas, & O. Westlund (Eds.), *Critical incidents in journalism: Pivotal moments reshaping journalism around the world* (pp. 71–84). Routledge.

Vogler, D., & Eisenegger, M. (2020). CSR communication, corporate reputation, and the role of the news media as an agenda-setter in the digital age. *Business & Society, 60*, 1957.
Vogler, D., & Gisler, A. (2016). The effect of CSR on the media reputation of the Swiss banking industry before and after the financial crisis 2008. *uwf. UmweltWirtschaftsForum, 24*, 201–206.
Wenzel, C. (2012). *Selbstorganisation und Public Value. Externe Regulierung des öffentlich–rechtlichen Rundfunks*. Springer.
Wilenius, M., & Malmelin, N. (2009). Towards sustainably managed media organizations: Reflections on the future of responsible business in media industry. *Business Strategy Series, 10*, 129–138.

# The Corporate Social Responsibility of the Media and the Turkish Media: Perspectives of Journalism Educators on Media and Corporate Responsibility

**Tevhide Serra Gorpe and Burcu Oksuz**

## 1 Introduction

Today, media organizations—as for-profit organizations—conduct corporate social responsibility (CSR) projects. In this way, they contribute to society and strengthen and/or establish relationships with their stakeholders. One of the reasons why they engage in CSR is related to the impact of CSR on their reputation, and media organizations are aware of this outcome; they are also in "business." Media institutions reach out to the public traditionally and on digital platforms to expand the reach of their social influence. Being able to use their own media channels is an advantage from their end because they can create awareness and make the public more informed and involved with the CSR projects they initiate. Media organizations have the communication power, which can help them spread their CSR projects easily. Along with this business perspective is the discussion of media responsibility (journalistic responsibility) and how they work together.

This study aims to explore the concept of media responsibility and CSR of media institutions in Turkey from journalism educators' point of view. It also aims to understand journalism educators' perceptions of *responsibility* (also from the business perspective) of the Turkish media. The research has practical implications on journalism education because it investigates how and to what extent these issues are integrated into curricula, including the challenges in teaching media and corporate responsibility to journalism students.

T. S. Gorpe (✉)
University of Sharjah, Sharjah, United Arab Emirates
e-mail: tgorpe@sharjah.ac.ae

B. Oksuz
Izmir Katip Celebi University, Izmir, Turkey
e-mail: burcu.oksuz@ikcu.edu.tr

 
F. Weder et al. (eds.), *CSR Communication in the Media*, CSR, Sustainability, Ethics & Governance, https://doi.org/10.1007/978-3-031-18976-0_12

The chapter reviews the media system, ethics, and self-regulation in Turkey, along with the CSR conceptualization in the media sector.

Data are collected from journalism educators who are working full-time at the journalism departments of faculties of communication in Turkey. A total of 12 semi-structured interviews are conducted. In this exploratory study, the data analysis is performed thematically. The main conclusion is that journalism educators do not support the idea of CSR for media institutions because they think that it should be inherent in the function of media institutions and within the individual responsibilities of journalists. Moreover, given the current condition of mainstream media, they do not think that media institutions care about CSR in a strategic and systematic way and even if they engage, the educators do not think it is performed in a genuine way. They suggest on how media institutions can demonstrate their social responsibility to the public. As indicated by the participants, media is the media of all publics, and the concept of "stakeholders" is not even an appropriate term to use within this context. They see issues with teaching media ethics and CSR to journalism/communications students and suggest how it should be done.

The research has significance in several aspects. First, only a few studies have investigated CSR in the Turkish media. Even for the journalism academics in our sample—who were former journalists in different media institutions, such as newspapers, news agencies, and televisions or served on journalism ethics committees—studying CSR in the Turkish media industry from a business perspective is a novel topic. Second, their perceptions of media responsibility, business responsibility, and the linkages they form are an eye-opener within the realities of the current Turkish media context. Journalism educators about the ideal and the reality. The research explores what is considered *ideal* from this perspective and what is the situation at present. Finally, the research carries practical implications for journalism education in Turkey. For example, İrvan (2006, p. 191) argues that students enrolled in ethics courses should be able to recognize ethical problem areas, be able to differentiate between law and ethics, and to know that legal actions are unnecessarily ethical actions. The role of an ethics course should be to help students with their analytical abilities and prepare them for the journalism profession.

The Turkish press has always been politically connected (Kaya & Çakmur, 2010). To understand Turkey's media system, one must go back and study the broader context, which includes Turkey's political, economic, and cultural histories, including its statism, nationalism, and secularism (Yesil, 2016). The analysis conducted for Turkey on the four dimensions of the media systems of Hallin and Mancini (structure and development of media markets, political parallelism, development of the journalism profession, and degree of state intervention) concludes that Turkey falls under the polarized pluralist model (Uce & De Swert, 2010). Bek (2011, p. 173) states that Turkey is similar to South and Latin American media systems, which are characterized by low circulation and high TV watching public broadcasting and regulation.

Bek (2011) states that compared with Southern European countries, one can say that through the establishment of ethical codes and the Press Council, the professionalization of journalism started early, but these initiatives and implementations were unsatisfactory to overcome the state and neoliberal market.

Yüksel (2013) argues that addressing the question of which media systems Turkey belong to is difficult because the Turkish media represents all the four theories of Siebert et al. (1963) that talk about public interest; some claim that they are supporting the interests of the state, whereas some state that they consider public interest; all variations can be observed in the media.

# 2 Literature Review

## 2.1 *Media Sector and CSR*

Corporate scandals, the increasing numbers of formal and informal protocols on corporate behavior, social issues, and public debates increase pressures on businesses to act socially. Media industries are no exception to this trend, although they lag in other industries in terms of CSR engagement and performance (Gulyás, 2009). As a result of neoliberalism, "public service areas, such as health, education, telecommunications, and mass communication, are opened to market competition and with the withdrawal of the state from social and cultural service areas, media organizations have to redefine their social legitimacy" (Taş, 2010, p. 3). Recently, media institutions' social responsibility has gained a more "systemic" view, where they are expected to inform stakeholders further on their issues (CSR in the Media Sector Guide, 2014).

Media institutions are themselves businesses, and this factor obviously has implications not only for the profitability drive perspective but also for broader corporate responsibility consequences. Similar to any other businesses, media institutions must identify their most significant environmental, social, and governance impacts (Grayson, 2009, p. 14). Unlike other for-profit contemporary (globalized) industries, the media has another important role in democratic societies: to provide basic information and analysis and to connect governments with their constituents, that is, to serve public interest (Lauk & Kuś, 2012, p. 168). Media institutions have a social function and given that they are the most important information provider to citizens, they have social responsibility (Ingenhoff & Koelling, 2012, p. 2). As stated by Grayson (2009, p. 3), "Media businesses have a profound impact on markets and societies through their abilities to inform, educate, influence, and entertain; and through their own content and the advertising they carry, media institutions have a pervasive influence on the global patterns of consumption." The difference between journalism and other types of professions has to do with the particular type of product or service professional journalism provides (Jacquette, 2016, p. 2).

The growing interest in the social responsibility of the media is not only related to the developments in its academic reflection but also to the structural trends within the media sector itself that form its empirical basis (Bardoel & d'Haenens, 2004, p. 166). Arriving at a satisfactory conceptualization of CSR in media industries is a difficult task for business ethics and communication studies (Painter-Morland & Deslandes, 2017, p. 665). As stated by Hou and Reber (2011, p. 166), CSR practices and CSR communication in other sectors have been studied by scholars, but in terms of media—as one of the influential industries—little research has been performed. "Social responsibility is interpreted in terms of 'responsibility,' indicating the media's responsibility with regard to society, and 'responsiveness,' suggesting the manner in which the media listens to and considers the public" (Bardoel & d'Haenens, 2004, p. 170). The main social responsibility of media institutions is to provide the necessary resources to produce journalism. Considering media and journalism as structurally and functionally different entities, a dichotomy between economic goals and social responsibility appears: Media brings profit, and journalism is ethical (Sandoval, 2013, p. 44).

The financial realities of today's media environment require media organizations to meet public interest and the financial expectations of their shareholders. Considering the special status of media companies in society, determining whether such companies integrate their CSR strategies into their strategic public relations management is important (Hou & Reber, 2011, p. 166). Han et al. (2008) state that in South Korea, with a few exceptions, most newspaper companies struggle to budget and invest thousands of millions of funds needed to demonstrate their social responsibility by donating and volunteering.

## 2.2 *Media Ethics and Self-Regulation*

Media ethics can be interpreted into a broader social responsibility term within the context of sociology or legal terms (Middleton, 2009). While journalists are ultimately responsible as individuals, they are responsible as individuals in an environment where their powers and duties are defined at least in part by their role in the corporate organization. The ethical dilemmas faced by journalists are numerous, and perhaps most often the conflict is between the role of the journalist as an individual and his/her employer's pursuit of profit (Richards, 2004, p. 119).

Professional journalism ethics serves an important function by helping journalists articulate and remind them of the moral choices they may face in writing their news; it makes them aware of their responsibilities and explain what is expected of them while they are accomplishing their duties. Ethics codes provide valuable information about what journalists perceive to be within their limits when they collect, organize, and report news (Jacquette, 2016, p. 3). In a study conducted by Basmacı (2021) on the ethical principles of 37 different internet newspapers, 10 newspapers included ethical principles, whereas 27 did not.

The evaluation of self-regulatory instruments is largely based on how they are applied in a particular country. If these instruments are established and if media self-regulation is part of the broad infrastructure, then journalists tend to rely more on these instruments than in countries where such practices are weakly institutionalized (Eberwein & Porlezza, 2014, p. 434). National and international press councils are one of the main self-regulation practices. These councils, which are established with the active participation of media workers and their free will, are self-control mechanisms that work to ensure a sense of social responsibility in the media (Arvas, 2011, p. 83). "The main function of press councils is to be a balancing factor between society and communication institutions. They act as an intermediary between the mass media and the masses to which these tools are addressed" (Girgin, 2007, p. 49).

Ombudsperson system is effective and important because readers are encouraged to have a say in the content of newspapers; in this way, the transparency of news organizations is ensured (Avşar, 2014, p. 59). "Giving regular accounts of the successes and failures of ombudsmen's outlets, they have a great potential of generating and keeping up public attention on media critical issues" (Lauk & Kuś, 2012, p. 171).

## 2.3 *Media Regulations in Turkey*

Several laws regulate the communication field in Turkey. These laws are the Press Law, Law on the Establishment of Radio and Television Enterprises and their Media Services, Radio and Television Law of Turkey, and relevant provisions of the Turkish Penal Code. The new Press Law, which was adopted in 2004, is one of the important regulations in the European Union Harmonization Process. It resulted in bringing changes to the old media law, which has been effective since 1950. The concept of Freedom of the Press, which was mentioned in the 1982 Constitution of the Republic of Turkey, was included for the first time to regulate the media field (Banar, 2006, p. 131).

Aydeniz (2018, p. 308) argues that the initiatives of media professionals are the most important sources for media ethics. The first initiative of the institutionalization of professional organizations is the attempt to establish the first civil professional organization of Ottoman journalists, which started with the proclamation of the Second Constitutional Monarchy and emerged on July 30, 1908, under the name Cemiyet-i Matbuat-ı Osmaniye (Aydeniz, 2018, p. 308). The First Press Congress was held on May 25–27, 1935. It was the first event in the Republic of Turkey that gathered journalists, politicians, the media, and the state (Atatürk Ansiklopedisi, 2022). The Cultural Commission was established in the Congress. It came up with propositions, new definitions, and criticisms about the press and the journalism profession (Arvas, 2010, p. 72).

The Congress was important in several dimensions. It helped develop professional cooperation, took initiatives to recognize the journalism profession as reputable, and mentioned the need for journalism education for the first time. Moreover, issues about journalism can now be directly transferred to government representatives for discussing solutions together, and the idea of a national journalism profession was born for the first time (Atatürk Ansiklopedisi, 2022). At the Press Congress, establishing a Press Chamber was decided (Arvas, 2010, p. 73).

The first text regulating the working relations of journalists in Turkey is the Press Union Law No. 3511, which was enacted in 1938 with the aim of institutionalizing the journalism profession (Üstün, 2013, p. 149). This law regulates the working conditions and self-control mechanisms of the journalism profession and makes it compulsory for journalists to be members of the Union (Kösedağ, 2018, p. 504). The Turkish Press Chamber was established in 1939, with the goal of institutionalizing the journalism profession in Turkey (Üstün, 2013, p. 150). It operated between 1938 and 1946. It was a compulsory professional organization established by law. Practicing journalism was impossible if they were not members of this organization. The media people had to work under a single roof, affiliated with a compulsory professional organization (Özgen, 2018, p. 134). "According to the Press Union Law No. 3511, the owners of newspapers and magazines published in Turkey and those who work in writing, news, pictures, photographs, and editing on a paid, regular basis and take up these jobs as a profession are obliged to become members of the Union, and journalism is prohibited outside the Union" (Baykal, 2021, p. 1959).

The Turkish Press Chamber was able to accommodate little of what was expected from it, and with a few exceptions, it could not fulfill its task of regulating the working conditions of journalists. Until 1946, when it was shut down, it continued to exist as professional police, which mostly kept the records of journalists, coordinated the relationship between the government and the media, and could impose sanctions if needed (Üstün, 2020, p. 42).

> The main objectives of the association were to protect the interests of its members, to protect the honor and dignity of the profession, to maintain order in the profession, to strengthen the feelings of solidarity, to open journalism schools or vocational courses, to make international contacts on professional issues, to represent the Turkish media, to keep the media alert and united for national interests, and to keep it loyal to the interests of the Republic. (Baykal, 2021, p. 1960).

With the closure of the Turkish Press Chamber in 1946, solid actions were taken in international relations toward the demands of the Press Freedom Committee of the American Society of News Editors and the United Nations press freedom principles (Üstün, 2020, p. 42). The Commission on the Freedom of Press, headed by Robert Maynand Hutchins, came to Turkey in 1945 and met with media representatives. This visit was important in terms of bringing media ethics into the agenda of the Turkish media as a topic to be covered in the Journalism Institute, which started at Istanbul University in 1950, and in monitoring the traces of media ethics in Turkey (Aydeniz, 2018, p. 309–310).

Turkish Journalists' Association was established on June 10, 1946 (Baykal, 2021, p. 1963). It is the oldest and most important professional organization in the Turkish media field, performing important functions since 1946 (Özgen, 2018, p. 119). In the bylaws of the Association, an effort has to be made to protect the principles of the profession. The Press Ethics Law was prepared under the leadership of the Association on July 24, 1960, and was signed by journalists and representatives of broadcasting organizations. A Press Court of Honour was established for the regulation of practices (Turkish Journalists' Association, tgc.org.tr, 2022).

The Press Court of Honour is Turkey's first voluntary press self-regulation organization (Baykal, 2021, p. 1963). Thus, the auto-control system in the media was established for the first time in Turkey in 1960. It was created to cover the fact that the Turkish press had a more limited role than the press council. It was an institution with a moral sanction that could only handle complaints about the press and make decisions about these complaints (Aydoğan, 1993, p. 188). The Press Court of Honour created a distance for journalists from the state, and the judiciary is important in terms of showing the relationship between the self-regulation activity and the state in the Turkish press (Baykal, 2021, p. 1963).

The Press Ethics Law agreed upon by journalists is not a law but has its foundation in the loyalty to the word given (Baykal, 2021, p. 1963). However, the media and journalists openly declared that they consent to the audit with the undertakings they signed (Özgen, 2018, p. 161). The enforcement power of the Press Court of Honour regarding self-regulation is limited. It cannot impose sanctions such as admonition, advice, warning, condemnation, and refutation against the complained media organ (Kösedağ, 2018, p. 504). The exposure was expected to be a sufficient sanction, as it would discredit the newspapers that violated the law (Baykal, 2021, p. 1964).

The Press Court of Honour carried out successful practices in the early days, and most newspapers meticulously attempted to comply with the law. However, it did not last long. Although political and social factors were considered effective in this condition, the inadequacy of moral sanctions was the most important reason (Uzun, 2009, p. 64). Due to the weakening of this mechanism, the Press Advertisement Institution was activated as a precaution (Kösedağ, 2018, p. 504). As a result of the application of Journalists' Union of İstanbul to the General Assembly of the Press Advertisement Institution, the advertisements of newspapers and magazines, which were given three warnings and one reprimand by the Press Honorary Council, were stopped by the Press Advertisement Institution. The intervention to cut income sources caused discomfort in media organizations, and they won the annulment case they filed with the Council of State. Media organizations also started to withdraw their commitments to the Press Court of Honour (Uzun, 2009, p. 64–65). Baykal (2021, p. 1964) mentions two basic issues related to it. First, journalists did not comply with the rules they promised to comply with and that the Court lost its control over the media. Second, a voluntary organization changed its status without the consent of the members. Özgen (2018, p. 166) states that the most important reason why journalists do not adopt the Court is that it has only a punitive nature and that they do not take initiatives to protect the rights of media members or to improve their personal rights. The Press Court of Honour remained dysfunctional and abolished itself in 1967 (Aydeniz, 2018, p. 310).

The Press Advertisement Institution was established on January 2, 1961, within the scope of the institutionalization efforts of the state in the distribution of its official announcements and advertisements to the print media (Üstün, 2010, p. 70). The main task of the Press Advertisement Institution is to announce official announcements and advertisements (Yıldırım & Kılıç, 2019, p 134). The Press Court of Honour based the sanction of cutting the advertisements of those who did not comply with the Press Ethics Law in cooperation with the Press Advertisement Institution (Arvas, 2010, p. 117). The Press Advertisement Institution, with the aim of bringing some national regulations, put the General Assembly Decision on the Ethics of Press consisting of seven articles into practice in 1994. Journalism will not be used for self-interest and immoral purposes, no distortions can be made in the news, and unproven news cannot be published are some of these articles (Banar, 2006, p. 132).

At the General Assembly of the Turkish Journalists' Association held on February 14, 1972, a nine-item text, previously prepared by the International Press Institute, was accepted as the "Ethics Codes of Journalism" (Turkish Journalists' Association, tgc.org.tr, 2022). In 1975, the Second Press Congress was held in Istanbul. In the Congress, the problems of the press and the implementation of ethical principles were discussed (Özgen, 2018, p. 177).

The Press Council was established in Turkey in 1986 to prevent restrictions on the media, upon the possibility of the political power to exercise further control over the media with the military intervention in 1980 (Uzun, 2009, p. 55–65). The idea of establishing the Press Council came about as a result of 28 journalists coming together in May 1986 to discuss and find solutions to the problems in media (Kösedağ, 2018, p. 505). For the establishment of the Press Council, the concept of self-regulation was thought to be the best practice and that the previous restrictions imposed by the media were insufficient (Şimşek, 2001, p. 299).

When the Council was established, a text called the Media Professional Principles, to which the members of the profession should be bound, and the Press Council Agreement were opened for the approval of journalists. Participation in the Council was based on volunteerism and took place with the acceptance of these two documents by the employee who would become a member (Özgen, 2018, p. 180). In the first few years after its establishment, the Press Council was considered an institution that is respected by the media and the Turkish society and whose decisions attributed importance. In the following periods, some member journalists reacted when decisions were taken against them and quit their membership (Kösedağ, 2018, p. 505).

Toward the 2000s, the Press Council was insufficient in terms of self-regulation. During this period, the Turkish Journalists' Association conducted studies on ethical principles: Turkey Journalism Rights and Responsibilities Declaration was published in 1998 to ensure the professional supervision of its members and to increase the quality of the journalism profession (Ayten, 2013, p. 6). Turkey Journalism Rights and Responsibilities Declaration was accepted by the Turkish Journalists' Association General Assembly on November 18, 1998, and was approved by its members at the General Assembly on March 31, 2000. The declaration was last updated on April 17, 2019 (Turkish Journalists' Association, tgc.org.tr, 2022). The Press Senate and the Professional Principles Monitoring Committee were established to monitor and control the applicability of the principles in this text, which are the most comprehensive and widely accepted professional principles (Aydeniz, 2018, p. 312). Turkish Journalists' Association is an important professional organization, and their declaration is determining the ethical codes of a specialized profession (Kaya & Uçak, 2020, p. 883).

The practice of ombudsperson or reader representation is not a well-known and practiced form of self-control in Turkey. An important start was made with the reader representative page launched in Milliyet newspaper on March 22, 1999 (Özgen, 2018, p. 196). Avşar (2014, p. 64) states that the first ombudsman practices in the Turkish media were made in the early 1980s by Seyfettin Turhan at Hürriyet News Agency and Emre Kongar at Hürriyet Newspaper. Seyfettin Turhan drew attention to the mistakes in news writing in the monthly bulletin he prepared for Hürriyet News Agency and reminded the basic journalism rules (Avşar, 2014, p. 64).

In the model of ombudsmen applied in Turkey, internal self-regulation is performed. Every media organization voluntarily appoints its own ombudsperson. The ombudsperson evaluates reader complaints about the media and comments and reminds journalists of their professional responsibilities within the institution (Özgen, 2003, p. 31). This model is criticized on the grounds that an ombudsman's salary is paid by the media institution he evaluates, and this situation may affect independence (Uzun, 2009, p. 54).

## 3 Methodology

The study examines the scope and frame of CSR in the Turkish media industry from journalism educators' points of view. Another aim is to investigate how CSR and ethics should be embedded in journalism education. The study is guided by the following research questions (RQs):

RQ1: What is the framework of CSR in media industries?
RQ2: What is the framework of CSR in the Turkish media sector?
RQ3: In what ways media ethics is linked with the CSR of media institutions?
RQ4: How should journalism educators/academics incorporate media ethics and CSR in curricula?

Given that research that looks at media industries as business entities and explores their CSR is limited, the present study used a qualitative approach and is exploratory in nature. The population of interest was journalism academics whose specific research areas are in press management/economics and media ethics. The sample comprised informed participants about the subject matter. The websites of the communication faculties of Turkey's Journalism departments were checked for accurate profiles of journalism educators who could take part in the study. The investigation started with one journalism academic who is personally known by one of the researchers. Purposive and snowball sampling methods were used. Snowball sampling is where the participants refer the researcher to other participants who can contribute to the study; this technique is considered effective for building up a sample quickly (Denscombe, 2013, p. 37).

**Table 1** The sample profile

| Participant | Gender | Years of experience | Rank |
|---|---|---|---|
| A | M | 19 | Dr. |
| G | M | 6 | Dr. |
| I | M | 23 | Associate Professor |
| M | M | | Prof. Dr. |
| S | M | 33 | Prof. Dr. |
| C | F | 11 | Dr. |
| S | F | 5 | Dr. |
| O | F | 10 | Dr. |
| Ba | M | 16 | Associate Professor |
| E | M | 3 | Dr. |
| B | F | 8 | Dr. |
| R | M | 21 | Dr. |

A total of 12 interviews were conducted with journalism educators in Turkey. The participants are from three cities in Turkey, which are Istanbul, Izmir, and Adana. The lowest rank of the participants was assistant professor. All the participants have a PhD in Journalism and out of the 12 participants, seven of them have professional media experience. The minimum academic experience of the participants was three years, and the maximum was 33 years. Among the 12 participants, four are female, and eight are male. Table 1 summarizes the profiles of the participants.

Semi-structured interviews have a form of an interview guide consisting of general questions, but the interviewers can vary the sequence of questions and can ask further questions (Bryman, 2016, p. 201). In this research, a total of 12 questions were asked to the respondents. Except for two interviews, the researchers conducted the rest together. One interview took place on Google Meet, 10 interviews on Zoom, and one was a phone interview. The interviews were conducted for over three weeks. The researchers sent an interview request email to each participant along with the objective of the study. The contact information of the academics was easy to obtain, either from participants' institution websites or from those who referred to potential informants. All the interviews were recorded with the consent of the participants. Before each interview, the researchers explained the aim of the study and informed the participants on the confidentiality and anonymity of the research. When the researchers asked for permission to record, most participants stated that doing so would be much better for them. During one interview, a technical issue arose; hence, the last part of the interview could not be recorded. The interview questions started from a general context and then moved to specific questions. The three main groupings of the interview guide were on CSR concept and media sector's CSR frame, the CSR of Turkish media organizations, and CSR and media ethics education for journalism students. The recordings were transcribed and during the interviews, both researchers took notes. Each researcher went over the data individually and came out with themes, trends, and conclusions. After this process, they discussed their individual work and finalized the themes and patterns, including which verbatim quotes from the participants should be included.

# 4 Results

## *4.1 CSR and CSR of Media Industries*

Most of the participants interviewed believe in the practice of CSR within the corporate world. They see it as a contribution to the society at large and a means of bringing reputation to the corporation. However, at the same time, they question the sincerity of these efforts. "CSR is instrumental" (Participant C). "I am in the mining sector, and I am doing a campaign for cleaning the sea" (Participant R). Several participants are skeptical about the CSR of the media within the context of Turkey. They consider the scope of CSR projects, and whether these projects being long term or not make difference in this respect. For example, CSR projects on social media are reported to reach a wide audience, but they mention some drawbacks: Being an activist should not be defined as someone who signs petitions but should be more than that. "...This is like forwarding blood donations on WhatsApp groups. It assumes that a duty is completed even without donating blood. Social responsibility is similar to this" (Participant M). CSR practices lack strategic components. Participants suggest that sectors and companies should run CSR projects with multisector involvement and that this is a teamwork.

The reasons for the CSR skepticism of media industries are multiple. First, if the media has performed its job properly, media industries no longer need to engage in CSR because it is already in their mission. "CSR and journalism benefit the society. In the journalism profession, public interest in involved" (Participant A). Second, the CSR concept within mainstream media is criticized. Mainstream media is underlined to be not functioning in alignment with journalism principles and ethical codes. Therefore, journalism educators do not really believe that the media sector can genuinely engage in CSR projects. As stated by Participant A, "Some things are happening now that are in conflict with the responsibility of the media to the public." The informants claim that mainstream media is under financial difficulties, and CSR is not and cannot be under its priorities. As stated by a journalism educator, "They fall into their troubles, they do not know how they can survive. As a result, the quality and number of CSR projects are affected" (Participant S). Journalism educators' retrieval of current CSR campaigns is limited, given the fact that almost no major CSR media campaigns are currently undergoing. The ones remembered are educational campaigns for girls, campaigns against violence to women and domestic violence, environmental campaigns, and campaigns with fundraising and support to disasters.

own media outlets. In this way, the media is seen as the communicator of CSR projects, but with a further involvement.

Third, media must make public service announcements. Specifically, it communicates or supports societal problems through direct messaging.

Fourth, media institutions should choose a CSR area where they can implement campaigns/projects similar to the practices of for-profit institutions in other industries. In this type of practice, the prioritization of issues is considered to be difficult. Media has a responsibility not to specific "stakeholder" groups but to the "public," and differentiating these issues in terms of salience is difficult. The participants do not believe in the strategic perspective of media CSR initiatives, and they consider the media responsible for all the issues in the society. They agree that if the media has done its job the way it should have, extra projects and campaigns are no longer needed.

Fifth, the media has a responsibility to touch upon issues of public interest. The way the media reports these issues should be in alignment with the ethics of reporting, such as accuracy, honesty, and impartiality. However, a contradiction is observed. The media can be sensitive to violence targeting women or domestic violence. It carries out campaigns, but the way the stories are reported are not aligned with what the campaigns stand for. This factor is exhibited in the headlines chosen, the visuals accompanying the stories, and the language of the news, creating the opposite effect. Another issue is linked with mainstream media being restricted to say what it has to say. Even if mainstream media is supportive of environmental issues, it is not articulate in situations that it assumes to be good for itself.

Sixth, the media must alert the public, engage in some sort of philanthropy, and push for volunteerism in times of crisis situations, such as natural disasters. This suggestion is made from the social aid perspective.

The participants discuss alternative media from the CSR perspective. Digitalization has created a myriad of people on the Internet, acting as journalists. Journalists who work in mainstream media leave their own media channels open in the digital world. Journalism educators state that journalism ethical codes are increasingly lacking, and journalists and other individuals who provide news in the digital world are operating with their own agendas. Sometimes, these agendas are restricted in mainstream media. One specific alternative media is mentioned as CSR being embedded in its system.

Social media is currently seen as the "news," and this perspective contradicts with the traditional view of journalism. Although the CSR projects of media institutions are viewed as having communicative power on digital platforms, the participants indicate that they are showcased as shallow projects, which people are aware of but do not have in-depth knowledge about.

The CSR areas for the media sector are mentioned as anything that falls under societal issues, issues that are of interest to the public. In addition to this macro-level perspective, journalism educators mention some specific areas of social responsibility, such as environment, human rights, and education in Turkey. Aydın Doğan Young Communicators Competition is cited as an exemplary project targeting communication students. Organizing summer camps for the youth from Anatolia and refugee children is regarded as an important societal issue. “All children should be opened to their dream world. The future is theirs” (Participant B.) Projects about women are also specified.

CSR priorities are youth, women, environment, human rights, education, refugees (children), disabled children, diversity, and literacy (media and digital).

The CSR perspective of media institutions has not been perceived as strategic. Rather than initiating CSR campaigns, the media sector can do a better job of tackling these social areas by covering the issues within the editorial context. In their view, a series of reports on an issue and its coverage in the media outlets can create awareness of the issue and can fit better with the journalism profession mission.

## *4.3 Media Ethics and CSR of Media Institutions*

Journalism ethics must be practiced to be able to talk about media organizations’ corporate responsibility. As stated by one participant, “If the focal point is not public interest, then talking about CSR is impossible” (Participant G). The discomfort of all the participants is that the journalism ethical codes are not applied. They highlight this concern with examples from a longitudinal perspective of the Turkish media industry development. “The biggest responsibility of a journalist is to ask questions” (Participant O). Participant G adds, “Reality, truth, and neutrality are basic principles, and these codes remain on paper. . .The freedom limits of the media are under the drawn limits of the powerful.” The participants also mention that in the Internet environment, the ethical codes are even looser or even nonexistent.

## *4.4 CSR/Media Ethics and Journalism Education*

The final research questions seek to understand how journalism educators inform journalism students on CSR and media ethics. One suggestion is that ethics should be covered in depth during the first year by courses in philosophy, sociology, ethics, and moral reasoning. More practical aspects can be delivered later. One participant reveals that foundation (private) universities are interested in attracting students; thus, program and course names are one way to do it. In this respect, they do not regard ethics as one of the critical topics. The participants also mention that studying ethics for one term is not enough. The attributes of the media professional teaching ethics courses are discussed. The professional experience is seen as helpful, but the educator also needs to know how to educate. Industry people should be selected carefully. The personal ethics of the industry professional teaching media ethics course is underlined. A media professional who has ethical violations in his/her work should not be teaching ethics to students even though he/she is a celebrity. The participants state that CSR is taught as campaigns in public relations programs. One participant has made a strong comment on who should *not* be teaching CSR. "CSR has to be the most innocent of all. A person with an advertising background should not lecture on CSR to students. . .People who are well-informed on CSR, experts, those who follow daily practices, and those from independent and from nonprofit organizations should teach CSR" (Participant Ba).

Within this context, issues with journalism education are raised. The participants argue that journalism education at faculties of communication should not be offered in a technical way, similar to vocational schools. They claim that the programs have content issues, and with the changing media landscape, changes must be done. Moreover, the faculties of communication and journalism programs in Turkey are more than needed. Journalism graduates have unemployment issues, which increase as time passes.

## 5 Conclusion

Almost all participants mentioned that the responsibility of the media is public interest. Doing their jobs, writing news objectively, asking the right questions, and following the story development are described among the responsibilities of journalists and media institutions. Journalism educators do not have many expectations from media institutions regarding their engagement in CSR initiatives. The role of media institutions is not to be the originator of CSR projects but more so to cover these issues in a series format within an editorial context and to support the CSR projects of other corporations by providing them in their media channels. The educators also suggest that media institutions should partner with other nongovernmental organizations and entities for CSR initiatives, rather than implementing them on their own.

The participants raise negative points about media organizations' social responsibility and ethics. They link this point to media ownership and the financial difficulties of media institutions. That is, these factors have an impact on media sectors' CSR practices. For example, newspaper circulations are going down, and the financial difficulties of the media sector—as it is the case in the world too—result in a change in the main goals of the media. They are concerned with survival; thus, advertorials and practices similar to advertorials are increasing in the media, and advertisers' influence on the editorial content is being observed. The participants all agree that advertising and editorial contexts are blurred now, which is a serious threat to the journalism profession.

Another point underlined by the educators is that journalists must pay attention to what they learn in news reporting when they are covering their stories. Journalism educators think that the way some news are covered is not reflecting the responsibility of journalists. News about women, children, and refugees have been criticized due to the language, visuals, and headlines accompanying these stories. The participants suggest that journalists should be informed through workshops and seminars, and self-regulation mechanisms must be activated.

The ethics codes of journalists must also be put into action, and these codes should be revised and updated. The educators argue that a huge difference currently exists between journalism education and the journalism practiced. Journalism education must be restructured, and ethics codes should not only be memorized but also internalized by practice. The participants do not think that there should be a separate CSR course in journalism curricula because the professional responsibility of journalists is overarching. Therefore, ethics should be part of all the courses in journalism programs.

## 6 Limitations of the Study and Future Research

The research involves journalism academics/educators. In the future, performing the same study with journalists who are working in the sector and then comparing the findings between the two groups will be helpful. In addition to qualitative studies, conducting a survey with journalism academics and journalists from different types of media outlets can also provide further information. Including managers of media institutions in future studies is equally important, so that people can be informed about their perspectives on running for-profit media organizations and how they approach the CSR of the media sector in Turkey.

## References


Arvas, İ. S. (2010). *Cumhuriyet Döneminde Basında Etik Bağlamda Ortaya Konulan Uygulamalar ve Bir Meslek Örgütü: Basın Konseyi.* İstanbul Üniversitesi Gazetecilik Doktora Tezi.

Arvas, İ. S. (2011). Basında Sosyal Sorumluk Anlayışı Bağlamında Günümüz Türkiye'sindeki Özdenetim Uygulamaları. *Akademia Erciyes İletişim Dergisi, 2*(2), 82–92.

Atatürk Ansiklopedisi (2022). *Birinci Basın Kurultayı* (25–27 Mayıs 1935). https://ataturkansiklopedisi.gov.tr/bilgi/birinci-basin-kurultayi-25-27-mayis-1935/

Avşar, Z. (2014). Özdenetim, Sosyal Sorumluluk ve Medyada Ombudsmanlık Uygulamaları. *Ombudsman Akademik, 1*, 59–68.

Aydeniz, H. (2018). Öz Denetim İmkânı Olarak Medya Ahlakı. In N. Erdoğmuş, Ö. Torlak, & K. B. Tiryaki (Eds.), *Temelleri ve Uygulamalarıyla İş Ahlâkı* (pp. 296–325). İlke Yayınları.

Aydoğan, F. (1993). Basın Otokontrolünde Basın Konseylerinin İşlevi. *Marmara İletişim Dergisi, 3*(3), 107–122.

Ayten, A. (2013). Bir Özdenetim Aracı Medya Etik Konseyi. *İstanbul Üniversitesi İletişim Fakültesi Dergisi, 44*, 1–17.

Banar, S. (2006). Türkiye'de Haber Verme İşlevinin Kişilik Hakları ve Etik Yaklaşımla Değerlendirilmesi. *Dumlupınar Üniversitesi Sosyal Bilimler Dergisi, 15*, 125–138.

Bardoel, J., & d'Haenens, L. (2004). Media meet the citizen: Beyond market mechanisms and government regulations. *European Journal of Communication, 19*(2), 165–194.

Basmacı, G. (2021). Examining internet newspapers based on ethical principles. In *1st International media and cultural studies conference abstract proceedings*, 6–7 September 2021, Gaziantep-Turkey.

Baykal, K. C. (2021). Türk Basın Tarihinde Özdenetim Kuruluşları İle Devlet Arasındaki İlişkiler. *Selçuk İletişim, 14*(4), 1948–1971.
Bryman, A. (2016). *Social research methods*. Oxford University Press.
CSR in the Media Sector Guide. (2014). Forumrsemédias, Orse. https://www.orse.org/fichier.
Denscombe, M. (2013). *The good research guide for small-scale social research projects*. Open University Press. McGraw-Hill Education.
Eberwein, T., & Porlezza, C. (2014). The missing link: Online media accountability practices and their implications for European media policy. *Journal of Information Policy, 4*, 421–443.
Gencel Bek, M. (2011). Turkish and international journalists comparing the media systems, journalistic practices and news production processes. *Medijska Istraživanja, 17*(1–2), 163–176.
Girgin, A. (2007). Yazılı İletişimde Meslek İlkeleri ve Basın Konseyi. *İstanbul Üniversitesi İletişim Fakültesi Dergisi, 27*, 47–62.
Grayson, D. (2009). *Corporate responsibility and the media*. Doughty Centre for Corporate Responsibility (Cranfield School of Management). www.doughtycentre.info
Gulyás, Á. (2009). Corporate social responsibility in the British media industries—Preliminary findings. *Media, Culture & Society, 31*(4), 657–668.
Han, E. K., Lee, D. H., & Khang, H. (2008). Influential factors of the social responsibility of newspaper corporations in South Korea. *Journal of Business Ethics, 82*(3), 667–680.
Hou, J., & Reber, B. H. (2011). Dimensions of disclosures: Corporate social responsibility (CSR) reporting by media companies. *Public Relations Review, 37*(2), 166–168.
Ingenhoff, D., & Koelling, A. M. (2012). Media governance and corporate social responsibility of media organizations: An international comparison. *Business Ethics: A European Review, 21*(2), 154–167.
İrvan, S. (2006). Academic approaches to ethical issues in "journalism ethics" courses. International symposium on media and ethics, November 3–4, 2006, Ankara, 187–204.
Jacquette, D. (2016). *Journalistic ethics: Moral responsibility in the media*. Routledge.
Kaya, R., & Çakmur, B. (2010). Politics and the mass media in Turkey. *Turkish Studies, 11*(4), 521–537.
Kaya, S., & Uçak, O. (2020). Gazetecilik Etik Kodlarının Oluşturulmasında ve Güncellenmesinde Meslek Kuruluşlarının Önemi: Türkiye Gazeteciler Cemiyeti Örneği. *Adıyaman Üniversitesi Sosyal Bilimler Enstitüsü Dergisi, 13*(36), 881–917.
Kösedağ, M. S. (2018). Yazılı Basında Öz Denetim Sorunu ve Basın İlan Kurumu'nun Denetim İşlevinin Basın Özgürlüğü Açısından Değerlendirilmesi. *Electronic Turkish Studies, 13*(10), 497–513.
Lauk, E., & Kuś, M. (2012). Editors' introduction: Media accountability—Between tradition and innovation. *Central European Journal of Communication, 5*(2), 168–174.
Middleton, M. (2009). *Social responsibility in the media*. http://www.mediafutureweek.nl/wp-content/uploads/2014/05/SR_media1.pdf
Özgen, M. (2003). Basında Bir Özdenetim Uygulaması Olarak Ombudsman. *İstanbul Üniversitesi İletişim Fakültesi Dergisi, 16*, 27–34.
Özgen, M. (2018). *Medya Etiği*. İstanbul Üniversitesi Açık ve Uzaktan Eğitim Fakültesi Ders Kitabı.
Painter-Morland, M., & Deslandes, G. (2017). Reconceptualizing CSR in the media industry as relational accountability. *Journal of Business Ethics, 143*(4), 665–679.
Richards, I. (2004). Stakeholders versus shareholders: Journalism, business, and ethics. *Journal of Mass Media Ethics, 19*(2), 119–129.
Sandoval, M. (2013). Corporate social (Ir) responsibility in media and communication industries. *Javnost - The Public: Journal of the European Institute for Communication and Culture, 20*(3), 39–57.
Siebert, F. S., Peterson, T., & Schramm, W. (1963). *Four theories of the press. The authoritarian, libertarian, social responsibility, and Soviet communist concepts of what the press should be and do*. University of Illinois Press.
Şimşek, M. C. (2001). Medya etiği. *Marmara İletişim Dergisi, 11*, 291–309.

Taş, O. (2010). Medya Etiğinin Tarihsel Temelleri ve Gelişimi. In B. Çaplı & H. Tuncel (Eds.), *Televizyon Haberciliğinde Etik* (pp. 3–23). Fersa Yayınevi.

Türkiye Gazeteciler Cemiyeti. (2022). *Türkiye Gazetecileri Hak ve Sorumluluk Bildirgesi*. https://www.tgc.org.tr/bildirgeler/turkiye-gazetecilik-hak-ve-sorumluluk-bildirgesi.html.

Uce, V., & De Swert, K. (2010). Introducing Turkey to the three media system models: The content of TV news in eleven countries. In B. Dobek-Ostrowska, M. Głowacki, K. Jakubowicz, & M. Sükösd (Eds.), *Comparative media systems: European and global perspectives* (pp. 63–76). Central European University Press.

Üstün, H. (2010). *Türkiye'de Devletin Yazılı Basına İlan ve Reklâm Dağıtım Politikaları ve Basın İlan Kurumu Örneği*. İstanbul Üniversitesi.

Üstün, H. (2013). Türk Basın Birliği'nin 1939–1946 Arasında Gazetecilerin Çalışma İlişkilerini Kurumlaştırma Girişimleri ve Günümüze Bıraktığı Miras. *İstanbul Üniversitesi İletişim Fakültesi Dergisi, 45*, 149–170.

Üstün, H. (2020). Türk Basın Birliği: Islah Girişimi ve Kapatılması (1944-1946). *İletişim Kuram ve Araştırma Dergisi, 50*, 28–45.

Uzun, R. (2009). *İletişim Etiği: Sorunlar Sorumluluklar*. Dipnot Yayınları.

Yesil, B. (2016). *Media in new Turkey: The origins of an authoritarian neoliberal state*. University of Illinois Press.

Yıldırım, G., & Kılıç, B. (2019). Basın İlan Kurumunun Yenilenen İmajının Hedef Kitleler Üzerindeki Etkisinin Ölçülmesi. *İletişim Çalışmaları Dergisi, 5*(2), 127–162.

Yüksel, E. (2013). *Türk Basını Hangi Kurama Uyar*. http://medyavehabercilik.blogspot.com/2013/10/turk-basini-hangi-kurama-uyar.html

# "Social Responsibility" as a Weapon?

## Public Service Journalism as Private Media's Main CSR Tool in Chavista Venezuela

**Elena Block**

> *Everyone sees what you seem to be, but few feel what you are...*
> *Machiavelli (2009, p. 66)*

## 1 Introduction: The Context

This chapter critically explores the viability of private news media's corporate social responsibility (CSR) programs in contemporary Venezuela, where deep political polarization, the mediatization of politics, and politicization of the news media have led to permanent antagonism between the authoritarian populist Chavista government and prominent private news media outlets. Among the casualties of the ongoing government–media clash are weakened, divided, or terminated traditional opposition parties and, more relevant to this chapter, a continuously battered and cornered journalism. The winner so far has been the Chavista clique which despite its national and international discredit and widespread accusations of corruption, inefficiency, and disrespect for civil rights, media, and democratic freedoms, remains entrenched in power by constructing what Lugo-Ocando and Cañizález (2021) analyzed as the "media cesarism" established by "digital age caudillos" (p. 90).

This state of affairs raises an important question: How can the waning Venezuelan private news media industry fulfil "social responsibility" functions and roles, operate legally, and also practice sustainable, public service, and engaged journalism? To answer this question, this chapter develops three lines of discussion: (a) It first explains the precarious situation of the Venezuelan news media ecosystem and overall political communication environment. It specifically discussed the Chavista government's understanding of the term social responsibility; (b) then, the chapter

E. Block (✉)
Strategic Communication at the School of Communication and Arts, The University of Queensland, St Lucia, Australia
e-mail: e.block@uq.edu.au



F. Weder et al. (eds.), *CSR Communication in the Media*, CSR, Sustainability, Ethics & Governance, https://doi.org/10.1007/978-3-031-18976-0_13

situates scholarly perspectives of the term "social responsibility" with a special section on CSR in Latin America; (c) after, there is an exploration of the links between CSR and civic and engaged journalism; and finally, (c) the chapter showcases the perspectives on CSR journalism of two Venezuelan news media editors.

For the purposes of this study, two semi-structured interviews were conducted with two very different types of editors[1]: Miguel Henrique Otero (personal communication, February 27, 2021), CEO and editor of one of the main legacy newspapers, *El Nacional (EN),* whose last print edition was published in 2018 and is now totally produced and delivered in digital form; and Luz Mely Reyes (personal communication, March 21, 2021), award-winning director and co-founder of *digital native* news media outlet *Efecto Cocuyo (EF).*

## 2 "Social Responsibility" as a Weapon for Media Intimidation and Censorship

The late President Hugo Chávez's government proposed and sanctioned the media's "Law on Social Responsibility in Radio, Television and Electronic Media" (2005, 2011) (the so-called *Ley Resorte or Resorte Law*) in 2004. This instrument was later refined and extended to digital media platforms in December 2010, and further reinforced during Nicolás Maduro's government in 2017 through the "Constitutional Law Against Hatred, for Peaceful Coexistence and Tolerance." Thus, instead of focusing on promoting democratic coexistence and more socially responsible media and journalism—as the name of the *Resorte Law* suggests—these legal instruments focused on regulating, censoring, and penalizing content and programming that were critical or offensive to the government, its policy and leaders.

A report by Freedom House (2015) indicated that the "*Resorte* Law" (2005, 2011) included severe restrictions that were "vaguely worded" and would potentially limit freedom of expression. The report stressed that "the law bans content that could 'incite or promote hatred,' 'foment citizens' anxiety or alter public order,' 'disrespect authorities,' 'encourage assassinations,' or 'constitute war propaganda'" (Freedom House, 2015, para. 4). The law embodies a sensitive approach to criticism of Maduro's inefficiency to handle frequent shortages of basic services, such as water and power, and basic food.

> Alleging that "democratic security" is at stake, the National Telecommunications Commission (CONATEL) uses the Resorte Law to stifle critical coverage of national and

[1]Questionnaires were sent to Otero and Reyes in January 2021. Otero responded via email and WhatsApp's audio. A Zoom recorded interview was conducted with Reyes. Their responses were freely translated from Spanish to English. Both respondents provided their written consent to use their names and content in this chapter. Their responses and consent are stored digitally in the author's personal and work devices. Dr. Franzisca Weder applied for ethical clearance for these interviews which was obtained from The University of Queensland for this and other academic projects.

international news, partly by imposing heavy fines on private television and print media. Many outlets have responded by softening their reporting. (Freedom House, 2015, para. 3)

Thus, the term "social responsibility of the media" turned into a dangerous penalizing weapon in the hands of the populist Chavista regime. This claim is demonstrated by the significant number of (a) media organizations (legacy, digital, broadcast) that were intimidated and penalized; (b) journalists who were persecuted, arrested, and/or imprisoned without due process; and (c) hundreds of news outlets expropriated or closed. Hugo Chávez until 2012, and his successor Nicolas Maduro until the moment of writing this chapter, have boosted an ongoing confrontational relationship with the private news media and their journalists. Chavismo also made what I called a "brutal" (Block, 2015) use of the media through two main strategies: (a) The overuse of public media channels, spaces, and regulations through which pro-regime content and overlong speeches that have been imposed to audiences and all private media outlets, which are legally compelled to transmit such content; and (b) political and legal control over the private media that still operates in the country through intimidation, threats, fines, and criminal penalization. Self-censorship has become not only normalized but a matter of survival for media outlets that decided to physically operate in the country, particularly TV channels such as Venevision, Televen, or Direct TV (national networks and cable, respectively). The cartoon below by Venezuelan visual artist Eduardo Sanabria (@EDOIlustrado 2013) represents the Chavista control over the media as a scary mediatic octopus whose tentacles represent TV screens (Fig. 1).

In his book, "20 Años de Censura en Venezuela," Andrés Cañizalez (2019, p. 10) documented the ways in which press freedoms have been undermined in the country, grouping them into six types of measures taken by Chavismo, namely,

(a) Direct closure of news media outlets framing them as "administrative measures," particularly in the case of radio and TV networks, such as popular pioneering TV network RCTV, a fierce and unapologetic Chávez's opponent, whose operating licence was not renewed by Chávez in 2007 and hence controversially expropriated and removed from the mediatic space.
(b) Obstruction of print media through the monopoly exercised by the government in the import of paper and other supplies.
(c) Direct violence against journalists to stop them from covering protests and demonstrations.
(d) Implicit or explicit censorship applied to some media prompting them to hide or fire "uncomfortable journalists" under threat of sanctions, fines, or closure.
(e) The use of a government-driven judicial system to apply pressure or thwart editorial postures under threat of penal sanction or prison.
(f) The sale of expropriated media to opaque capitals or figureheads of the regime.

These measures, together with the creation and sanctioning of legal instruments, such as the "*Resorte* Law" (2005, 2011), as well as Hugo Chávez's compulsory, overlong divisive and confrontational broadcast messages, progressively helped Chavismo take over the country's mediatic spaces. During his 13 years in power,

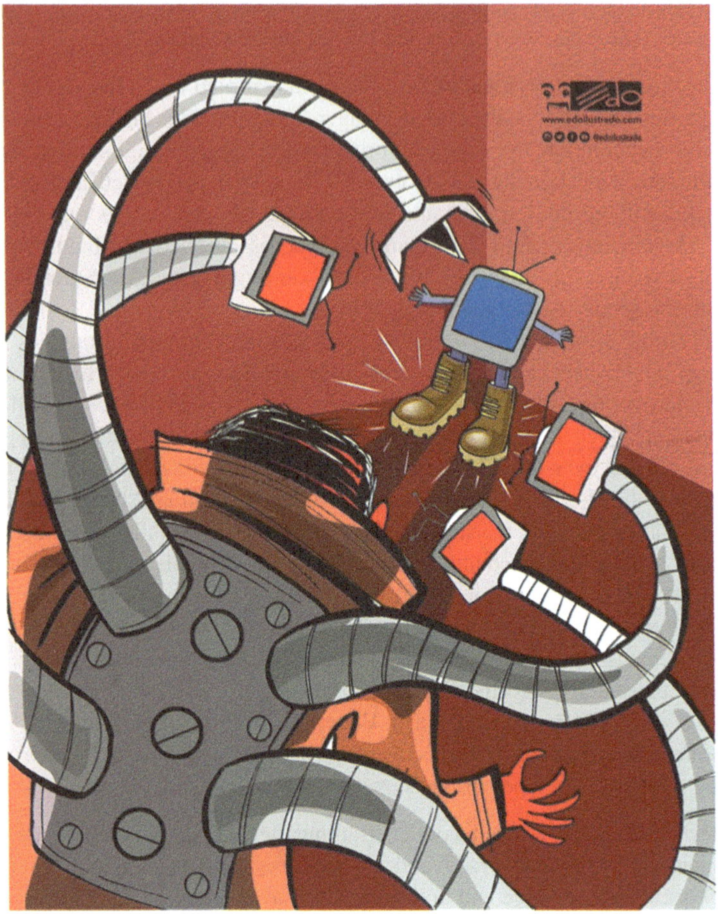

**Fig. 1** "Chávez Octopus" (2013). Courtesy of Venezuelan contemporary visual artist Eduardo Sanabria (EDOIlustrado)

Hugo Chávez built what Bisbal (2009) called "communicational hegemony," or what Lugo-Ocando and Cañizález (2021) defined recently as "media cesarism."

Social responsibility programs voluntarily enacted by Venezuelan media companies have had two avenues of action. The first was social programs that focused mainly on education, media literacy, civic education (*EN*, Cisneros organization, *EC*, Televen); culture and arts (*EN*); and on children's health, education, and food programs targeting poor communities (mainly by private or commercial TV networks and cable TV such as the Cisneros Organization, Direct TV, and Televen). The second was through practicing what my two respondents, *EN*'s Otero and *EC*'s Luz Mely Reyes, interpret as public service journalism, which will be discussed in the last section. But first, the next sections will give an overview of scholarly approaches to CSR and public service and engaged journalism.

## 3 A Communicative Approach to CSR

Machiavelli (2009) suggested half a millennium ago (see the vignette that starts this chapter) in a visionary representation of modern public opinion, *The Prince*—interpreted here as individual or collective political, social, religious, or corporate leaders—must not only have good qualities, but "it is most necessary for him to appear to have them" (p. 65). CSR is the discourse and strategy used by organizations not only to build good character, qualities, and deeds but also to *appear* to have them for public opinion purposes via *communication*, as the opinion of 'the many' matters.

Thus, CSR embodies a strategic communication discourse, and practice based on philanthropic deeds enacted by affluent members of a group or society. CSR has been loosely described as a *social contract* of sorts between businesses, organizations, stakeholders, and the society in which they operate and thrive. From the patronage of the arts by ancient *Maecenas*, or the noblesse or largesse of some feudal lords, to the philanthropic philosophy of some British and American industry revolution magnates, social responsibility discourses have been underpinned by ideologies advocating the moral obligation of the rich to help the least privileged (see Heath, 2013). While the first factories were rapidly growing, nineteenth century critics pointed to harsh working conditions for all, especially women and children, which was leading to increasing poverty and social unrest. Some industrial improvements and welfare movements started to emerge. Since then, the discourse of social responsibility slowly started to take shape, to be finally situated, honed, and ultimately normalized from the 1950s onward, when Howard R. Bowen and Johnson (1953) suggested that business should take a wider social responsibility in society beyond profit and gain.

Weder and Karmasin (2016), who have developed a rich body of work on the subject, advanced a critical communication approach to CSR triggered by the interrogation of the role of business and its contribution to society. Their argument centers on addressing a key question, "how to communicate about CSR and communicate responsibly at the same time" (Weder & Karmasin, 2016, p. 72). This is not a trivial question as it shows that "CSR communication" has a crucial role to play in any organization. These authors viewed CSR communication in relational terms, as the commitment and responsibility assumed between an organization and its stakeholders. *Sustainability*, Weder and Karmasin (2016) suggested, is the "character" or ethos of the *responsible relationship* that develops from an integrated approach to CSR communication; in this context, these authors consider two dimensions of communication: (a) As "a way of taking responsibility" (p. 77), and (b) "communication management as responsibility" (p. 77), which for them is a must-have value within a responsible and ethical vision of strategic communication.

Thus, from Weder and Karmasin's perspective (2016) organizational strategic planning of activities should aim to work on a set of crucial CSR topics, namely,

- The integration of social and ecological aspects into business activities.
- The voluntariness and self-commitment beyond legal regulations.

- The orientation toward stakeholders.
- The creation of an added value for society and companies.
- The acceptance of responsibility (in social, ecological, and economic respects).
- (p. 74).

Moreover, from a business public relations (PR) perspective, Page and Parnell (2021) defined CSR as the "purposeful incorporation of public interest into business decision making" (p. 200), an interpretation based on what Carroll (2015) called a "triple bottom-line," or three Ps, namely, people (social), the planet (environment) and, of course, profit (economic), that emerged from the more socially conscious 1960s society. Drawing on Christensen et al. (2007), Page and Parnell (2021) added that by following the CSR pathway, organizations can both address or pre-empt the economic, social, and environmental impacts that they exert on the community, as well as the concerns and expectations of their main stakeholders. Unlike Weder and Karmasin (2016), Page and Parnell (2021) do not include communication in their "pathway."

In sum, CSR has been traditionally viewed as the way businesses *give back* through philanthropic, good-will, or *doing-good* publicity and PR programs that fund and/or advocate for social causes, such as education, health, technology, food and farming, diversity and culture, and the environment. It could be implied then, that some business interpretations of CSR PR point to the need of building brand reputation and legitimacy.

Indeed, some critical or skeptical scholars and commentators have regarded CSR as a legitimizing tool used by organizations for spin, self-promotion, or greenwashing, as a communication strategy aimed at building, managing, and validating a goodwill reputation that guarantees their prestige and position of power (at the level in which they operate) in society. This critical view embodies a tension between some hypocritical or tone-deaf organizational campaigns and the increasing demands from today's hyper-engaged audiences for truly or "*authentic*" ethical industry practices, which include more effective and interactive listening, transparency, empathy, authenticity, and sustainable practices.

The resolution to this tension could be in part located in the arguments of critical communication scholars, such as Weder (2021), who has viewed CSR and its links with PR from the perspective of agonic dialogue, solution, and social change. By problematizing the issue of sustainability, Weder (2021) added the importance of incorporating "the plurality of agonistic voices in stakeholder engagement processes in social, cultural and environmental transformations" (p. 1) to the CSR discussion, in an interpretation that exceeds traditional philanthropy or *do-good* approaches.

Going back to Machiavelli's quotation, CSR is the way business leaders and corporations strive both to build a good character or ethos and also "appear" to have such a do-good character for public opinion purposes via *communication*. CSR is a legitimatizing communication and brand identity strategy and discourse with the potential to build trust between organizations and their stakeholders. Also, as Weder (2021) suggested, CSR can help to trigger plural dialogue and eventually social change. In this task, corporations and businesses of all sizes and types need all forms of media and news media, influencers, and social media commentators and reviewers, to provide third-party validation.

The foregoing discussion suggests that business organizations face a contemporary dilemma, suggested by Bachmann and Ingenhoff (2017) drawing on Bartlett (2011), in the following way: "companies 'are damned if they do [CSR communication] and damned if they don't'" (p. 80). This is especially the case in contemporary times, branded by the Edelman Trust Barometer (2021) as the era of "information bankruptcy," when post-truth misinformation reigns. In this context, both business organizations and the news media play significant roles in restoring trust and providing reliable information to their publics. For some, socially engaged and public service journalism can be the sustainable solution to media's CSR practice, especially in countries where democratic freedoms and basic goods and services are a luxury, such as Venezuela. This explains why this chapter takes Weder and Karmasin's (2016) and Weder's (2021) critical communication approach to CSR to analyze the situation in chavista Venezuela, where CSR programs might have the potential to mobilize public dialogue about citizens' urgent, agonic demands for the sustainable politics and policy that the country has lacked is the last two decades. But before reviewing CSR journalism in Venezuela, and as a way of context, a reflection on some of the approaches to CSR in Latin America.

# 4 CSR in Latin America

A study undertaken by The Economic Commission for Latin America (ECLAC—CEPAL as it is known in Spanish), written by Correa et al. (2004), suggested that business social responsibility has been a theme discussed in Latin America from the early decades of the twentieth century, usually approached from the perspective of philanthropic and charity exercises by private companies. Correa et al. (2004), explained that the 1990s saw a growth of academic and industry interest in the topic, as an important pillar of business management. Correa et al. (2004) found that most of the private CSR initiatives in Latin America then were voluntary, top down, and developed with little or no consultation with stakeholders. The topics addressed then reproduced or adjusted CSR topics discussed in the USA and Europe, instead of addressing local themes.

However, there were some significant initiatives that started immediately before or with the millennium: 118 companies in the region subscribed to the United Nations Global Compact (n.d.-a), officially launched by the UN in 2000, whose mission has been to "mobilize a global movement of sustainable companies and stakeholders to create the world we want" (United Nations Global Compact, n.d.-b). Previously, an organization called "Forum EMPRESA" (Crossroad Global Hand n. d.) was born as a hemispheric alliance of business organizations created in 1997 to promote ethical CSR in the continent. This Forum advocated for a clear vision that integrated respect for ethical values, people, communities, and the environment.

Since then, the importance of CSR has been normalized in Latin America, always present in the private and corporate sector's conscience. Education, environment, and responsible marketing have been among their main tasks in the region. In Venezuela, Víctor Guédez (2009) developed an important body of work that proposed a relational and dialogic approach to CSR focused on the importance of stakeholders, with a special focus on "interest groups." Guédez (2009) considered interest groups as the main trigger factor for organizational strategic planning. He advanced a model based on four principles, communication, inclusion, equity, and reciprocity. These principles, he suggested, should be effectively mobilized (and never sacrificed) by the corporate sector to ideally achieve harmony between the interests of the organization, its stakeholders, and the community.

Scholars from the University of Zulia researched CSR perspectives and practices during the first decade of the millennium. Romero (2010), for example, approached CSR in Venezuela from an ethical perspective, playing a significant role within an ethical model of development. Romero's (2010) views were inspired by the principles of the Global Compact (n.d.), which proposed a "holistic" model of business equipped to face the political, economic, and social challenges present in Latin America and Venezuela. However, Romero (2010) concluded that despite its merits, this approach was insufficient as it tended to reinforce the same neoliberal models and issues that it had initially tried to address or resolve.

## 5 CSR, Public Service, and Engaged Journalism in the Americas

The news media, be they legacy, alternative, or digital natives, have the double character and function of publishing CSR stories produced by other businesses and, at the same time, insofar as commercial companies themselves, private news media outlets also need to gain legitimacy and validate themselves as socially responsible outlets by incorporating CSR programs and strategies into their own PR management (Bachmann & Ingenhoff, 2017; Hou & Reber, 2011; Ruiz Malbarez & Tamayo Cevallos, 2019). This is one of the communicative spaces where journalism meets PR.

Specifically, from a Latin American perspective, Ruiz Malbarez and Tamayo Cevallos (2019) represented the news media as public, private, and community entities that provide a "mass communication service" (p. 1) through different distribution channels. These Latin American authors have showcased an implicit partnership between business CSR and the news media to offer programs that demonstrate socially responsible behavior or performance. In this partnership, the news media has a double role expressed in, namely (a) the impact that their content has on their publics and therefore in the formation of public opinion, which they call the "social mission" of the media; and, (b) in the fact that media are businesses themselves that need to be ethical, accountable and transparent about their own finances, governance, and social performance, and sustainable in relation to the environment. However, as Ruiz Malbarez and Tamayo Cevallos (2019) explained, news media companies tend to regard their social role more as "observers, translators, mediators and communicators" (p. 5) of socially responsible content produced by others rather than that of their own.

Hou and Reber (2011) explored CSR programs by major global media American companies that included CBS, Time Warner, Disney, and Comcast among others; their study revealed that the main CSR topics disclosed or publicly reported by these companies on their websites included environmental policies and activities, community relations, diversity, employee relations, human rights, and "Media specific CSR activities" (p. 168). The most mentioned were "providing responsible media content, committed diversity in content, and supporting creativity," with none of them disclosing "whether they have activities on media literacy or media education" (Hou & Reber, 2011, p. 168). It is interesting to note here that this last finding contradicts the main CSR activities carried out by the Venezuelan news media (before and during the first years of Hugo Chávez's rule) that mainly focused on education and media literacy, as the last section of this chapter will explain.

As Hurst and Ihlen (2018) suggested, CSR and the term "engagement" seem to go hand in hand in the literature in pieces where engagement has been regarded as an instrumental technique used by business organizations to not only "be" but to "appear" socially responsible. In their reflection about this topic, Hurst and Ihlen (2018) suggested that instead of focusing on "managing expectations" (p. 133), CSR and engagement should involve strategies aimed at the creation of "networks of mutual responsibility though relationships" (p. 133) that exceeds one-way communication as "organizations need to listen, adapt, and change" (p. 133) vis-à-vis all their stakeholders, which is consistent with Weder's (2021) agonic dialogic approach to CSR.

News media organizations are business entities in their own right and also need to engage with their stakeholders, a group that encompasses not only their journalists, employees, sources, partners, shareholders, press associations, and unions, but also, or especially, their audiences. Wenzel and Nelson (2020) explained engaged journalism in terms that evoke the definition of PR, as a way of managing relations of trust and loyalty between journalists and their publics. This definition could be interpreted in instrumental or transactional ways, as a reputation strategy tool. However, it could also be viewed as a dialogic, participatory, community-building strategy whereby community members can participate in shaping the news agenda. Wenzel and Nelson (2020) mentioned that some news outlets even hire "engagement editors, and reporters" (p. 515) that interact with audiences online and offline, mainly via social media (they mentioned Facebook), which rather than "cannibalize" their stories or ratings render "a clear positive relationships between the two" (p. 516).

In a similar vein, Carrie Brown (2019) provided a celebratory view of engaged or social journalism, which she described as "doing journalism with and not for the communities" (para. 2). Consistent with Wenzel and Nelson (2020), Brown (2019) also described this approach as not only listening to people's need but especially "working with the community to produce impactful journalism" (para. 3), in a process that she suggested has been viewed, at least in some towns and cities in the USA, as the emergence of journalistic jobs described as engagement or "community impact" reporters. This approach stems, of course, from the 1990s trends described as people powered, public, or civic journalism that originated in Jay Rosen's (1995) propositions whereby the news media become conversational spaces as for him "journalism cannot remain valuable unless public life remains viable. If public life is in trouble. . . .then journalism is in trouble" (p. 35). This is a view that becomes not only more relevant but urgent in times of conspiracy theories and the antidialogic, antagonistic style of Donald Trump (Block, forthcoming). Therefore, as Rosen (1995) implied "journalists should do what they can to support public life" (p. 35).

The view above ties in with the renaissance of what Harrison (2019, p. 1) calls "public service journalism," a "profoundly ethical" view of this practice, described in terms of an *ideal* place or "mediapolis," a communicative space "where information and opinion may be expressed civilly to enable good choices be made and public concerns to be thoughtfully addressed" (p. 1). This approach also conceives the news as dialogical rather than top-down purely informational and monological. A model that seems, perhaps too "ideal" if we think about the disruptive confrontational and impolite discourse and tone that have prevailed in mainstream and social media conversations in the last few years, where hate speech and cancel culture reign. Either publicly or privately funded, or both, what matters for Harrison (2019) is that public service journalism serves "democratic outcomes" by providing "conditions of coexistence" (p. 10) and trustworthy stories that embody shared experiences in today's complex and heterogeneous societies.

Focusing on digitally native news, Ferrucci (2015) advances from Rosen's (1995) views of public journalism, and purports what he branded as a "new form of public service journalism," equally focused on engagement but without the "negative" ideological connotation attached to Rosen's initial propositions because some critics transformed journalists into activists or advocates of democracy. For Ferrucci (2015) the new, digital native version of public service journalism rescues the notion of engagement as news organizations should aim to "to truly matter to the community it serves" (p. 907); Ferrucci (2015) understands engagement as communicating with readers and citizens and responding to the communities they serve, a task that has become easier and natural to non-profit digital native news outlets that are not governed by markets or editorial politics or ideologies. This discussion raises the question, is public service journalism the way contemporary journalism and news media outlets can enact an authentic, transparent, and trustworthy CSR practice? Is working with the communities they serve, listening to their issues and reporting on them, the way to provide socially responsible journalism? From different perspectives, this chapter's respondents believe that public service journalism is the answer, at least in embattled Venezuela.

## 6 Two Editors, Two Outlets, Two Perspectives

Why did this study purposively interview two completely different types of editors to illustrate and support this chapter's discussion? This Chapter's aim was to explore two "typical" cases that provide the main features and feasibility of CSR journalism in Venezuela. Both editors were very different in the following aspects:

(a) History and background: *EN* was founded on August 3, 1943, when Venezuela, a country used to caudillo power clashes and the succession of autocratic nineteenth-century dictators, was beginning to wake up to democracy. *EN* started as a quality, democracy driven, and intellectually led spreadsheet that became *the* learning place for young journalists; on the other hand, *EC* began in 2015, when the Maduro regime was increasing measures that rather than authoritarian were becoming dictatorial, dangerous, and increasingly threatening for journalists and opposing politicians; overcrowded prisons were seeing more and more political prisoners.
(b) Vision: *EN's* Miguel Henrique Otero is a legacy news media owner who comes from an inherited traditional turned digital media business model centered on advertising and now internet traffic; *EC*'s Luz Mely Reyes is a journalist co-founder of a digitally native news outlet based on a more contemporary business model, primarily based on crowdfunding, international cooperation funds, and donations.
(c) Audience: *EN* has a larger and broader mixed audience, while *EC* has a more community engaged, niche, stakeholder type of audience.

News content is released every day via their respective websites and distributed *transmedia* posted to the main social media platforms, distributed via WhatsApp and other private messaging platforms. For example, a pdf version of *EN* that nostalgically evokes, and actually reproduces, the front page and full printed paper, is directly delivered digitally to loyal sets of audiences, through databases accumulated by Otero personally, and his staff across the years. *EN* maintains the coverage of the main domestic, Latin American, and international news, digitally but still in legacy broadsheet style, with news and long-form covered and produced by a newsroom now reduced to a minimum. As used to be customary in the print version, now a digital editorial section showcases op-ed articles by intellectuals, academics, and political and economic experts. The *Papel Literario* (Literary Paper) is a weekly weekend section that develops critical analysis of Venezuelan literary and philosophical topics and publishes interviews with writers, poets, and philosophers. El Nacional is mainly free for digital readers who are regularly asked to make voluntary monetary contributions to maintain "free journalism."

Miguel Henrique Otero, President and CEO of *EN*, is an example of a traditional but tech-savvy media owner, who was a pioneer in migrating to digital. *EN* was published in both print and digital formats, until its print edition was terminated in 2018. Otero regularly engages with his audiences via WhatsApp and Twitter, where he has accumulated 1.8 million personal followers and follows 1 million, which is rare for this type of mega influencer that very rarely follows back (see @miguelhotero). *EN*'s (https://www.elnacional.com/) last printed edition was published on December 15, 2018, mainly due to government monopoly and restrictions on printing paper (see "Venezuela's El Nacional", 2018). The Associated Press ("Venezuela's El Nacional", 2018) reproduced the newspaper's statement on that occasion,

> El Nacional's front page delivered a defiant tone, telling readers that becoming an exclusively online publication wouldn't change its core mission of holding the powerful accountable—especially at this moment in Venezuela's history. "El Nacional is a warrior and will continue to fight". (paras. 3-4)
>
> ...the paper's president and CEO Miguel Otero expressed his regret at halting the printed version of a newspaper his grandfather and father headed. "They have not been able to do us in, nor will they ever," he said. (para. 8)

Since then, *EN* has become an exclusively online news outlet successfully transmediated across social and private messaging platforms. At the moment of writing this chapter, the paper had 5.3 million followers on Twitter (@ElNacionalWeb, 2021, September 6). Otero, personally, had 1.8 million followers. Although *EN* has been politically engaged in the past, and even supported Hugo Chavez in 1998, they never abandoned their watchdog style. This explains why Chávez started his attacks against the news media and specific journalists, like Otero himself, just over a year after being first elected in 1998. Chávez and now Maduro have been unable and/or unwilling to accept criticism. In 2020 The World Law Foundation (n.d.) awarded EN and its editor, Miguel Henrique Otero, for their fight for democracy and freedom of the press.

*EC* on the other hand, was born as an independent, community-engaged news media outlet, founded and carefully crafted by recognized Venezuelan journalists Luz Mely Reyes, Laura Weffer, and Josefina Ruggero, who by 2015 had been affected by the precarious situation of the press in Venezuela. *EC* explores issues that are socially important and often controversial in a predominantly Catholic Venezuela (topics that range from domestic violence, LGBTQ, and gender equality issues to abortion). *EC* is a non-profit, digital native, news media start-up that has accumulated 739.7 thousand followers on Twitter since its foundation in 2015 (see Efecto Cocuyo, 2021). Award-winning journalist, co-founder, and director, Luz Mely Reyes, personally has 207.1 thousand Twitter followers (see Luz Mely Reyes, 2021). Reyes (personal communication, March 21, 2021) is very proud of the fact that the outlet posted their first tweet on January 8, 2015, and in two days they gained 12 thousand followers.

*EC* covers Venezuelan main news but with a community focus; long-form reports on developments on COVID-19 and the general health crisis in Venezuela, stories about the increasing number of Venezuelan migrants, or on how a pineapple cost more than a minimum salary in a local market. *EC* was one of the first news outlets to produce an investigative piece denouncing illegal mining in the Orinoco River's *Arco Minero* that is destroying Venezuelan forests (see Pineda & Lopez, 2020). There is also a fact-check section about conspiracy theory-like stories or government lies that are an important part of their website content and headlines. There is a section on sexual health and teen-pregnancy education.

Both media outlets have been affected by governmental intimidation, threats, and censorship via the Resorte Law and other legal instruments. *EN* has been especially hit not only by penalties, fines, confiscation of assets, and censorship but also by physical and discursive attacks by Chavista leaders and their followers, or "*colectivos*." In 2021, a $13 million compensation defamation suit by Diosdado Cabello, one of the main heads of chavismo, culminated in the government seizing the newspaper's Caracas headquarters. Cabello, Chávez's former vice-president and president of the National Assembly, has been "obsessed" with *EN* after the newspaper published a report in 2015 that accused him of being the head of a drug trafficking military gang, called "*el Cartel de los Soles*" (see El Nacional, 2021). *The Washington Post* published an account of the impasse where they reproduced the accusation that a Chavista drug cartel has ruled Venezuela in the last decades, and where they also called *EN*, "the last independent newspaper" (Diehl, 2015, para. 6) operating in the country.

## 7 Consensus: Public Service Journalism as Private media's CSR Tool

The foregoing discussion suggests that both, *EN* and *EC*, regardless of their differences, view the work of investigative journalism as a public service and as a socially responsible and engaged practice, fully committed to cover and report on the issues that are relevant to their respective audiences. Both are focused on covering issues that matter, presented truthfully and professionally, with the aim to hold power accountable regardless of government intimidation. Their situation is so delicate that both Otero and Reyes now live and work in exile. Otero left in 2015—the same year that *EC* was born—when the threats became so real that he had to either leave or go to jail. Reyes also received threats as *EC* increased its reports on increasing restrictions to press freedom and journalists' arrests.

This leads to the main issue interrogated by this Chapter: What are, if any, El Nacional's and Efecto Cocuyo's CSR programs? How do Otero and Reyes see CSR applied to news media companies or enterprises today? Their answers to these questions respectively mark the different realities and journalistic approaches that they represent. However, they also reflect an agreement about the fact that the only way to develop an authentic CSR practice in Venezuela is through public service and community-engaged journalism.

Otero (personal communication, February 27, 2021) explained that the news media has a dual character because the news media can develop CSR programs like any other commercial outfit but with the added potential that it can multiply their impact and outreach via publication of their content or messages. However, he immediately added that under the current critical Venezuelan circumstances it is,

> impossible to think about traditional corporate social responsibility programs when the goal becomes to survive, either because of the economic crisis that pulverizes the company's income and because of the harassment of the regime.
> (Otero, personal communication, February 27, 2021)

However, *EN*'s focus is not only on denouncing the government's corruption and mishandling of the economic, social, and political crises but also on education, the environment, on health, and science, which demonstrate that *EN* still fulfils its engagement commitment with their audiences. In the past, the newspaper used to allocate budget resources and specific departments to engage with community issues at all levels. He added that in the past *EN* developed several CSR programs related to the educational area (mainly media literacy among primary and high school students). For example, "*El Nacional en el aula*" ("El Nacional in the classroom") was a program where the newspaper financed both editorial products and instructors to work in classrooms with young students, motivating them to read through the newspaper, even using competing newspapers. The company managed several initiatives to award creativity in creative writing, visual arts, and science, at the national level, which included the best original story, cartoons, and children's literature, among others.

*EN* was one of the first newspapers to publish a section on social media "trends," or "*tendencias,*" discussing topics attractive to their readers. Otero (personal communication, February 27, 2021) added that "El Nacional was considered a pioneer in supporting all online community issues."

Luz Mely Reyes (personal communication, March 21, 2021), started the interview by claiming that *EC* was born not as a newspaper but "as a social responsibility model." She added,

> Efecto Cocuyo was born in Petare [popular area in East Caracas]. Initially, we started doing a rather symbolic street fundraising that also had the aim to chat with people in the community; we started by asking people "What is not normal?" and they responded telling us their main issues. The hashtag #itisnotnormal (#noesnormal) became trendy in [a] short time. In two days we had 12 thousand followers on Twitter, everything organic, nothing promoted. We felt that something new was starting to happen. It was then that we defined our journalism as public service journalism, and we started to see ourselves as public servants.

Reyes (personal communication, March 21, 2021) indicated that this view about public service journalism was another point of difference vis-à-vis traditional journalism; she continued her argument by suggesting that,

> . . .legacy media represents itself as the only representative or owner of the truth and the agenda setters par excellence, focusing on certain topics for editorial purposes. We were not like that, for it is about the agenda of the people, the audience sets the agenda. . . .It is not about clickbait, it is not purely a monetary interest that focuses on audiences for profit purposes. . . .in a polarized country like Venezuela, our work is to listen to our audiences, to cover their stories with independence. The news agenda has to be consistent with the people's agenda. We do not see audiences as passive, like traditional media used to do. True, information and news may generate profit, but information is also a human right, we all have the right to be informed, which is a huge responsibility for us as journalists.

Reyes flagged then the disruptive situation triggered by the split between business models that have dominated the news media in the last decades. The progressive destruction of print media in Venezuela left many journalists unemployed. Reyes worked as a journalist and editor in traditional newspapers for decades and participated in the convergence process that took print media to multimedia and digital. She suggested that many newspapers did not fully accomplish this transition from the old business model to the new digital one. In addition, she remarked, some outlets remained loyal to their ad-centered model and to their top-down information-out ways instead of becoming more community and audience centered, adopting the new social paradigm where the *social* is embedded in the organization's vision, mission, and structure (see Kim, 2020).

Reyes also highlighted that newsrooms were, and in some cases still are, rather undemocratic, vertically conceived spaces where the editorial line reigned. Reyes explained that she and her partners, journalists Weffer, and Ruggiero, decided to build their outlet under the new social, community orientated, nonprofit digital native paradigm, where editorial decisions are taken by consensus, and values are marked by their love for a socially responsible, ethical, dialogic, and engaged journalism, consistent with the models of engaged, "agonic dialogue," and public service journalism proposed in the previous section by Carrie Brown (2019), Wenzel and Nelson (2020) and Weder (2021). Luz Mely Reyes won the Lasa (Latin American Studies Asociación—*Asociación de Estudios Latinoamericanos*) in 2021 for *EC*'s outstanding journalistic coverage and innovation model ("Crisis Global", 2021).

Reyes suggested that *EC* has two programs that could be called CSR, the first consists of regular community engagement meetings to listen to their audiences and monitor their issues and needs, called "*Guayoyo*[2] *con Cocuyo*" ("Having Coffee with Cocuyo" in English) that take place offline and now also online due to COVID-19 restrictions. The second is a journalistic education, or rather professional development program, called "Escuela Cocuyo" (*Cocuyo School*), a successful CSR initiative separate from the newsroom although some of the journalists may participate in the sessions. The last edition is about the coverage of Venezuelan migration, "from the inside and from the outside" ("Como cubrir", 2021, para 1), sponsored by the Deutsche Welle (DW) Akademie.

However, for Reyes, as for Otero, despite the media and journalistic education initiatives they have carried out, they see that their most important CSR work does not and cannot lie in traditional corporate "programs," but rather in their everyday work reporting the news; holding power accountable, providing trustworthy, ethical, community engaged, public service journalism, and in their daily coverage of stories that are important to the country, to their specific audiences and communities.

[2]Venezuelan slang for a light, rather watery, type of coffee.

## 8 Conclusion

The perspective of two different news media editors leading two very different private media outlets in Chavista Venezuela revealed important differences between legacy/commercial and digital native/crowdfunded journalistic practices, structure, and vision. However, their perspectives also revealed important coincidences about the way both types of news media outlets approach their CSR role in terms of *public service journalism*. They both see their role not only as primary providers of news but also of education and media literacy. They do not view their public service role as an image-driven reputation strategy but rather as a *public service duty* as media outlets that need to mobilize the public debate on key issues affecting Venezuela and its citizens in the context of an authoritarian populist regime. Otero (personal communication, February 27, 2021) argued that rather than a matter of reputation it is a matter of engaging with issues that are important to their audiences; he said that the newspaper aimed "to dynamize a discussion about various political and social issues of collective interest," such as, for example, the environment and sustainability, a topic that used to have a regular section for decades in the print version of EN. On the other hand, Reyes (personal communication, March 21, 2021) is very proud of the fact that their digital outlet emerged in a popular Caracas suburb (Petare) and was born as a news vehicle that aimed to *engage* with the community and their issues rather than imposing an editorial line. Both believe that their outlets mobilize public discussion about political, economic, and social issues and that, despite the intimidating media environment, both, in their own ways, contribute to hold power accountable.

## References

Bachmann, P., & Ingenhoff, D. (2017). How do media companies gain legitimacy? An experimental study on the (ir) relevance of CSR communication. *International Journal of Strategic Communication, 11*(1), 79–94.

Bartlett, J. (2011). Public relations and corporate social responsibility. In O. Ihlen, J. L. Bartlett, & S. May (Eds.), *The handbook of communication and corporate social responsibility* (pp. 67–86). Wiley.

Bisbal, M. (Ed.). (2009). *Hegemonía y control comunicacional*. Alfa UCAB.

Block, E. (2015). *Political communication and leadership: Mimetisation, Hugo Chávez and the construction of power and identity*. Routledge.

Block, E. (Forthcoming). *Discursive disruption, populist communication and democracy: The cases of Hugo Chávez and Donald J. Trump*. Routledge.

Bowen, H. R., & Johnson, F. E. (1953). Social responsibility of the businessman. .

Brown, C. (2019). *Engaged journalism: It's finally happening*. Nieman Lab. https://www.niemanlab.org/2019/12/engaged-journalism-its-finally-happening/

Cañizález, A. (2019). *20 años de censura en Venezuela*. Editorial Alfa.

Carroll, A. B. (2015). Corporate social responsibility. *Organizational Dynamics, 44*(2), 87–96.

Christensen, L. J., Peirce, E., Hartman, L. P., Hoffman, W. M., & Carrier, J. (2007). Ethics, CSR, and sustainability education in the financial times top 50 global business schools: Baseline data and future research directions. *Journal of Business Ethics, 73*(4), 347–368.
Como cubrir la inmigracion venezolana desde adentro y desde afuera (2021, n.d.). *Escuela Cocuyo*. https://escuelacocuyo.com/
Correa, M. E., Flynn, S., & Amit, A. (2004). *Responsabilidad social corporativa en América Latina: una visión empresarial*. Cepal.
Crisis Global. Desigualdades and Centralid de la Vida. (2021, n.d.). LASA 2021. https://lasaweb.org/es/lasa2021/lasa-media-award/
Diehl, J. (2015, May 24). A drug cartel's power in Venezuela. *The Washington Post*. https://www.washingtonpost.com/opinions/a-drug-cartels-power-in-venezuela/2015/05/24/9bc0ff14-ffd6-11e4-8b6c-0dcce21e223d_story.html
Edelman. (2021). *2021 Edelman Trust Barometer*. https://www.edelman.com/trust/2021-trust-barometer
Efecto Cocuyo [@EfectoCocuyo]. (2021). *Posts* [Twitter profile]. Retrieved September 5, 2021, from https://twitter.com/efectococuyo?lang=en
El Nacional Web [@ElNacionalWeb]. (n.d.). *Posts* [Twitter profile]. Retrieved September 4, 2021, from https://twitter.com/i/lists/73405501
El Nacional, la otra obsesión de Diosdado Cabello (2021, May 3). *El Nacional*. https://www.elnacional.com/venezuela/el-nacional-la-otra-obsesion-de-diosdado-cabello/
Crossroad Global Hand (n.d.). *Forum empresa*. https://www.globalhand.org/en/browse/csr/organisation/organisation/21455
Ferrucci, P. (2015). Public journalism no more: The digitally native news nonprofit and public service journalism. *Journalism, 16*(7), 904–919.
Guédez, V. (2009). *Los grupos de interés: de lo transaccional, a lo relacional ya lo consustancial*. Boletín CIES.
Harrison, J. (2019). Public service journalism. In J. Harrison (Ed.), *Oxford research encyclopedia of communication*. Oxford University Press. https://doi.org/10.1093/acrefore/9780190228613.013.867
Heath, R. L. (2013). Corporate social responsibility. In R. L. Heath (Ed.), *Encyclopedia of public relations* (Vol. 1, pp. 207–211). Sage.
Hou, J., & Reber, B. H. (2011). Dimensions of disclosures: Corporate social responsibility (CSR) reporting by media companies. *Public Relations Review, 37*(2), 166–168.
Hurst, B., & Ihlen, O. (2018). Corporate social responsibility and engagement: Commitment, mapping of responsibilities, and closing the loop. In K. A. Johnston & M. Taylor (Eds.), *The handbook of communication engagement* (pp. 133–148). Wiley.
Kim, C. M. (2020). *Social media campaigns: Strategies for public relations and marketing*. Routledge.
Ley de Responsabilidad Social en Radio y Televisión y Medios Electrónicos (2005; 2011). [Ley Resorte] Gaceta Oficial No. 38.333, December 12, 2005; February 7, 2011. Bolivarian Republic of Venezuela.
Luz Mely Reyes [@LuzMelyReyes]. (2021). *Posts* [Twitter profile]. Retrieved September 5, 2021, from https://twitter.com//luzmelyreyes?lang=en
Lugo-Ocando, J., & Cañizález, A. (2021). The return of the caudillos in the digital age–changing hegemony and media Caesarism: Continuities and changes in the news media landscape under the Chavismo. In E. Korin & P. Pain (Eds.), *When media succumbs to rising authoritarianism* (pp. 90–105). Routledge.
Machiavelli, N. (2009). *The prince*. (P. Constantine, Trans.). Vintage Books.
Page, J. T., & Parnell, L. J. (2021). *Introduction to public relations: Strategic, digital, and socially responsible communication*. Sage.
Pineda, J., & Lopez, E. (2020, July 15). Arco Minero del Orinoco: crimen, corrupción y cianuro. *Efecto Cocuyo*. https://efectococuyo.com/especial/arco-minero-del-orinoco-crimen-corrupcion-y-cianuro/

Rosen, J. (1995). Public journalism: A case for public scholarship. *Change: The Magazine of Higher Learning, 27*(3), 34–38.
Ruiz Malbarez, M. C., & Tamayo Cevallos, C. D. (2019). El doble rol de la responsabilidad social empresarial en los medios de comunicación. *Cofin Habana, 13*(suppl.1), e14.
Romero, N. (2010). La responsabilidad social corporativa:¿ una estrategia de cambio hacia un modelo ético de desarrollo? *Revista de Ciencias Sociales, 16*(3), 456–467.
Sanabria, E. [EDO Ilustrado], (2013, November 10). 'Chávez Octopus' [Cartoon]. Copyright [2013] by Eduardo Sanabria EDO Ilustrado. Reprinted with permission.
United Nations Global Compact (n.d.-a). Latin America Caribbean. https://www.unglobalcompact.org/engage-locally/latin-america
United Nations Global Compact (n.d.-b). Our Mission. https://www.unglobalcompact.org/what-is-gc/mission
Venezuela (2015, September 10). *Freedom House.* https://www.justice.gov/sites/default/files/pages/attachments/2015/09/10/fh_fop_2015_venezuela.pdf
Venezuela's El Nacional publishes last print edition. (2018, December 18). *Associated Press* (AP). https://apnews.com/article/1a49b15031d846a6b0751050c6de9fe3
Weder, F. (2021). Strategic problematization of sustainability reframing dissent in strategic communication for transformation. *Public Relations Inquiry.* https://doi.org/10.1177/2046147X211026857
Weder, F., & Karmasin, M. (2016). Communicating Responsibility: Responsible Communication. In Handbook of Integrated CSR Communication (pp. 71–86). Springer International Publishing. https://doi.org/10.1007/978-3-319-44700-1_5
Wenzel, A., & Nelson, J. L. (2020). "Engaged" journalism: Studying the news industry's changing relationship with the public. *Journalism Practice, 14*(1), 127–128. https://doi.org/10.1080/17512786.2019.1585121
World Law Foundation (n.d.). *The World Jurist Association awarded in Colombia Press freedom and democracy.* https://worldlawfoundation.org/2020/02/18/the-world-jurist-association-awarded-in-colombia-press-freedom-and-democracy/

# Subsistence Journalism: Corporate Control and Corporate Change in Queensland Regional Journalism

**Richard Murray**

## 1 Introduction

The story of the global news media over the past two decades is one of chaos and uncertainty. The well-documented digital disruption of journalism as an industry has raised existential questions that have gone largely unanswered (Picard, 2014; Villi & Picard, 2019). News media at all levels has been impacted the ceaseless changes in how news is funded, produced and disseminated. Globally, nowhere have these changes been more keenly felt than at the local level (Hess & Waller, 2016). Lauded as the grassroots of democracy (Pickard, 2019), local and regional journalism has declined to the point of collapse leaving only news deserts behind (Abernathy, 2016, 2020). Pickard (2019) demonstrates when local and regional news media disappears, the communities they served become less democratic and less sustainable as organs and subjects of democracy. In other words, without the press to perform a watchdog role, less scrupulous actors in society are free to do as they please free from public scrutiny. Abernathy (2020) observes a correlation between communities that have lost their local news media and a rise in corruption on the part of elected officials. A recent report from the Reuters Institute at The University of Oxford (Schulz, 2021) reveals this is far from a Western phenonium. The report says:

> Across the world we have seen a spate of cutbacks and layoffs, and longstanding titles closing down. This is troubling because local and regional news media can play a critical role in informing citizens and underpinning democratic processes.

The report also reveals the decline in local and regional media has accelerated during the COVID-19 pandemic, and this is particularly because this has been a period when people have needed reliable information the most.

R. Murray (✉)
Digital Journalism, The School of Communication and Arts, The University of Queensland, St Lucia, QLD, Australia
e-mail: r.murray1@uq.edu.au

 
F. Weder et al. (eds.), *CSR Communication in the Media*, CSR, Sustainability, Ethics & Governance, https://doi.org/10.1007/978-3-031-18976-0_14

The story of the decline in Australia's news media and journalism is increasingly well documented. The once-coveted rivers of advertising gold that nourished and sustained Australian journalism have dried up. What has been left behind is a fragmented and impoverished journalism industry. Along with thousands of job losses, hundreds of journals across Australia have closed or become shadows of their former selves. Nowhere is this more so than in Queensland.

The mainstream Australian news media is arguably one of the most concentrated of any plural democracy (Lidberg, 2019). Lidberg argues this concentration results in a media system that lacks a diversity of views and opinions and, therefore, represents the views of Australia's social and economic elite. This is a media system that is based in Australia's most populous and affluent city, Sydney, and it is largely focused on events in Sydney and New South Wales. More crucially, in the context of this study, this leaves large tracts of Australian news media dependent on the fortunes and whims of dominant news providers including Rupert Murdoch's News Corp and the Australian Government funded Australian Broadcast Corporation (ABC).

Over the past decade, both metro and regional journals in the state have been owned and controlled by Australia's largest news organization, News Corp. This paper investigates who and what filled the void left by News Corp journals in two regional towns in Queensland. Through interviews with journalists, some of whom have moved from News Corp newspapers to the new regional news publications, this study seeks to understand these changes through the eyes of those charged with reporting regional Queensland news.

## 2 Australian News Media in Decline

The recent story of Queensland's regional news media has been one of carnage. Hess and Waller (2016, 2021) have documented in detail both the decline of regional news in Australia, but also regional Australia as a site of innovation, change and rebirth in journalism. Researchers at Canberra University's News & Media Research Centre (Fisher et al., 2020; Park et al., 2021) have found that while Australian journalists tend to like their work, the pace of change in regional journalism makes imagining future innovation and employment difficult. Örnebring (2018) highlights the link between precarity of employment for journalists and a sense of stasis in relation to the news organisations who employ them. Downman and Murray (2017) argue that when change and innovation do come it is more likely to happen at the local and hyper-local levels of journalism. These organisations are smaller, require less start-up capital, and they speak to and for clearly defined audiences. Their small size also exposes them to a marketplace that is best described as Hobbesian: making life for many nasty, brutish and short.

The Australian Public Interest Journalism Initiative (PIJI) (2021) tracks the state of journalism across the country by monitoring news organisations' openings and closings. Their analysis shows the number of news and other journalism publications has been in steep decline over the past five years. In regional and remote areas, which often rely on a single publication for local news, the spectre of news deserts (Abernathy, 2016, 2020) looms large. Indeed, according to Wahl-Jorgensen, the 'greatest challenge facing journalism today lies in the collapse of local news provision' (2019: 163). This, in turn, has serious consequences for local communities and, more broadly, the health of democratic societies (Olsen et al., 2020; Wilczek, 2019). Of interest here is the assumption that local journalism and grassroots democracy are inextricably linked. As Deuze and Witschge (2020) argue in theorising a post-digital disrupted journalism, in many instances, journalism and democracy have already decoupled, and it is not always useful to assume journalism's contemporary role serves democracy. Papacharissi (2021) takes a different view arguing that in the interests of sustainable democratic communities journalist must act as "democracy's detectives" (p 128). She argues it is a journalist's duty to do so and not doing so threatens the health and fabric of democratic society. This view is in keeping with the philosophies of Milton (2014) and Mill (1989) and is at the heart of liberal-democratic values.

The term 'news deserts' was coined by American journalism academic Penelope Muse Abernathy (2016). Like PIJI in Australia after her, Abernathy tracked the decline in news organisations in the United States. She found that where local news coverage had disappeared, there had been a rise in public and private corruption and a decline in community cohesion.

## 3 Disrupted News in Queensland

In May 2020, much of Queensland looked to be on the verge of news desertification when—during Australia's first COVID-19 lockdown—the largest news organisation in Australia, the Rupert Murdoch-owned News Corp, announced it would be closing 14 regional newspapers and migrating a further 100 to digital delivery (Kelly, 2020). This came at the cost of an estimated 500 jobs. As a casualty of the digital disruption of journalism, News Corp's news division has been in decline for many years. Despite its drastic restructuring of the Australian regional news mastheads, and posting a corporation wide profit of US$330 million in the 2020–2021 financial year, revenue across News Corp's Australian news division fell by 21% (Butler, 2021).

A number of News Corp newspapers have only been able to survive in digital form. For journals such as the *Gympie Times*, the *Sunshine Coast Daily*, the *Noosa Times*, the *South Burnett Times* and Ipswich's *Queensland Times*, their news and other content can now be found on the Brisbane metro news daily, the *Courier Mail*, as a tab on its website. For their part, News Corp has heralded this move as a 'a significant digital opportunity' (News Corp Australia, 2020) for regional Queenslanders and the news they access. Whether this claim is true or not is a question for another day. But it certainly created an opportunity for the Victorian news organisation, Star News Group, to step in. At the time of writing, Star owned

32 newspapers across Victoria and Queensland (Star News Group, 2021). Ten are based in Queensland and all were founded in 2020 and 2021. The move also provided continuing professional opportunities to many of the displaced News Corps journalists. As Dodd and Ricketson (2021) report, for many journalists, life after lay-off is difficult with many unable to find other roles in journalism. Exactly how profitable this is for Star News Group is unknown. What is known is in less than a year, Star has filled a void left by News Corp to become too dominant player in regional Queensland news. This study focuses on the Queensland towns of Noosa and Gympie and the journalists who work there at the Star News Group owned *Noosa Today* and *Gympie Today*. This study focuses on the challenges journalists working under these mastheads face in getting a weekly edition to market, and what this means for the kinds of stories they cover for their respective communities.

Both publications launched weekly print editions in July 2020. *Noosa Today* prints 22,500 free copies a week with 5000 being distributed through local supermarkets and newsagents. The remainder is delivered directly to reader's letterboxes. *Gympie Today* prints 20,000 copies a week, and these are sold through local newsagents, supermarkets and petrol stations at three dollars per copy. *Gympie Today* also delivers directly to subscribers. Both publications also offer digital versions of the print editions. These are distributed via email and Facebook. The exact reach and impact of the digital editions are unknown.

In December 2020, in a break between COVID-19 lockdowns, this researcher spent several days with the editorial teams at *Noosa Today* and *Gympie Today* conducting ethnographic observation (Madden, 2017). These observations were enhanced by semi-structured group and individual interviews (Kvale, 2012) with journalists working as part of these editorial teams. What follows is based on the data gathered during this period of fieldwork.

### *3.1 Noosa*

Noosa is a resort town about two hours north of Queensland. Located at the northernmost tip of the Sunshine Coast, Noosa is famous for its beautiful beaches, national park and high property prices. With average prices hovering around the million-dollar mark and with houses in Noosaville and close to the beach regularly fetching more than five million dollars, Noosa is the most expensive place to buy in Queensland (Lamb, 2021). Noosa's identity is distinct from that of the neighbouring Sunshine Coast in both cultural and administrative terms. Noosa is its own shire giving the town independent local governance from the Sunshine Coast bordering them to the south and west. This has enabled Noosa Shire Council to control its shape, style and pace of development. It has also enabled the council to protect the national park that dominates the town. Together, these factors combine to create an exclusive enclave for Australia's elite and well-to-do.

### 3.1.1 Gympie

Fifty minutes drive to the northwest of Noosa is Gympie. The former gold town is a shadow of its former self although in the last two years property prices have spiked as the Sunshine Coast and Brisbane to the south have surged. Local attractions include relics from a golden past—a museum, a steam-powered locomotive and a Nestlé factory. For people passing through on their way north to the Wide Bay, the Capricornia Coast or further afield, and south to the Sunshine Coast, Brisbane or the Gold Coast, Gympie serves as a convenient place to stop and buy fuel or a meal from one of the many fast-food restaurants on the highway that snakes its way through the west of the town. Gympie also serves as a staging point for the towns of Tin Can Bay and Rainbow Beach on the Cooloola Coast to the east. The contrast with Noosa is stark.

## *3.2 The People on the Newspapers*

In keeping with North's (2009) findings on the gendered newsroom, both journals display similar gender-based staffing structures. At the apex of each organisation is a late middle-aged man in the twilight of his career. These men are employed on a part-time basis and to a large extent cover the kinds of stories they like. In Noosa, this man is Phil Jarrett. He is a renowned surfing journalist and writer, and much of his journalism for *Noosa Today* looks at the local surfing scene. He has an interesting journalism pedigree. He was the founding editor of the ironic *Tracks* surfing magazine before entering what he terms 'the flesh years'—a decade spent as editor of Australian *Playboy* magazine and then at the helm of *Penthouse* magazine. After that, he graduated to jobs in Europe and was for a time the CEO of surfing and snowboarding clothing company, Quiksilver Europe. Jarrett's laidback character and affluence are reflected in the tone and style of *Noosa Today* and a tendency toward what Hanaucsh terms 'lifestyle journalism' (2012, 2014).

At *Gympie Today*, the senior journalist is Arthur Gorrie, a traditional 'ink on the fingers' kind of journalist. Before reporting for *Gympie Today*, he worked his way up into the upper echelons of News Corp's *Gympie Times*. Here, he cut his teeth doing police rounds and reporting to the local courts. This set the tone for the rest of his career and his eventual move to *Gympie Today*. With Gorrie as part of the editorial team, *Gympie Today* has developed a reputation for taking on Gympie's rich, powerful and corrupt. This can be seen in Gorrie's work on the Fraser Island bushfires where he drew public attention to local and state government inaction in dealing with the fires resulting in a state government inquiry. Gorries said his watchdog journalism had endeared him to local and state authorities rather than creating distance. The paper also features plenty of court reporting.

Both publications have former News Corp journalists at the next level: middle-aged women employed on a part-time basis who complained their work took well over 40 hours per week. At *Noosa Today*, Margaret Maccoll fills the 'middle role' in the newsroom, covering the local courts and council rounds, but the bulk of her journalism is 'soft journalism'. Stories on local events and profiles of local people of interest dominate. Some of what Maccoll covers is news, but none of her journalism is investigative.

At *Gympie Today*, the woman in the middle is Donna Jones, a career journalist at News Corp's *Gympie Times* before its demise. Like Maccoll, she is employed on a 20-hour-a-week casual basis, yet Jones said she tended to work 40–60 hours per week. Jones specialises in local news and events covering everything from local children's sports to obituaries. Her work is down-to-earth and plays a big part in *Gympie Today's* community-focused tone and style. Jones described her work as a 'calling' and said that without her passion for journalism and the community she serves, maintaining such high work hours for such small pay would not make sense.

At the bottom of the newsroom staffing structure were young women: at *Noosa Today* Abbey Cannan and at *Gympie Today* Rose Astley. Both are graduates of the University of the Sunshine Coast's journalism programme. Both are the only full-time members of their respective editorial teams. Cannan's days are full at *Noosa Today* since, on top of reporting duties, she manages the newspaper's website and social media presence. Cannan said, like Maccoll and Jones, there were not enough paid hours in the week to get through her work and she often worked well above her paid hours. This raises interesting questions about the potential for burnout, the stresses of the job, and Cannan's expectations for longevity in journalism.

At *Gympie Today*, although things were better for Astley, they were still far from perfect. At the time of the interview, she had been working as a journalist for 18 months. She had started with the News Corp's *Gympie Times* months before losing her job in one of many restructures that had seen the paper gutted. She had found work at *Noosa Today* before settling at *Gympie Today* in her hometown. Astley said the big difference between her experience as a journalist at the *Gympie Times* and her work at the *Today* mastheads was that News Corp provided structured and systematic training which was not offered at the Star News Group.

Like Cannan, Astley said it was seldom she managed to stay under her 37.5 hours per week. Unlike Cannan, Astley did not have to busy herself with managing her publication's social media. This was not because someone else on the *Gympie Today* staff was handling this task. Rather, the editorial staff at *Gympie Today* had decided it was better to concentrate their efforts into their print product. This decision had proved prudent given the devasting impact Facebook's shut-out of Australian news had had on publications similar in size to *Gympie Today*. For Astley, the time she did not have to spend maintaining the newspaper's social media presence she was able to dedicate to telling stories from her community. Although Star Media did not offer formal professional development, Astley said she was supported and mentored by both Gorrie and Jones in her journalism. It was for this reason Astley could imagine a future in the industry. Gorrie, Jones and Astley were interviewed as a group and the affection and respect they had for each other were obvious.

## 4 Regional Media Conditions

The Star News Group offered conditions that made mainstream regional journalism in Queensland possible. The interviews revealed Star provides the production infrastructure, the masthead and a means of distribution that made *Noosa Today* and *Gympie Today* possible. These are the factors that make community news start-ups out of reach for many news entrepreneurs (Hess & Waller, 2016). In terms of infrastructure, Star Media provides a masthead and a logo (reminiscent of *USA Today*). They also provided the means of production. Aside from a physical newsroom equipped with the day-to-day reporting necessities, Star provided sub-editing and layout services. In line with the kinds of outsourcing and decentralised production practices that have become the norm in mainstream journalism (Downman & Murray, 2017), all the layout and sub-editing for both publications are done by an independent contractor in the Philippines. For the former News Corp staff interviewed, this was a point of concern since they missed their publications being produced under one roof.

In addition, both publications made the move from online to full physical print editions during the first months of the COVID-19 crisis. This coincided with News Corp largely pulling out of Queensland's regions. At the time of the interviews, all print editions for Star Media's Queensland mastheads were printed at a plant in New South Wales. This created logistical challenges in the timely and profitable distribution of print editions in their respective communities. This was remedied in June 2021 after Star News Group purchased a printing plant in the regional Queensland Centre of Rockhampton to service their Queensland publications (Pearce, 2021). The plant had been owned and operated by APN News and Media and the *Morning Bulletin*, a News Corp-owned Rockhampton newspaper that had abandoned print as part of News Corp's digital-only strategy. At the time of sale, the plant was wholly owned by News Corp, after APN had sold their New South Wales and Queensland regional holdings, referred to as Australian Regional Media (ARM), to News Corp in December 2016 (Australian Competition and Consumer Commission, 2016).

The acquisition by Star of the printing plant was important for it showed the newsgroup was prepared to make a significant investment in its regional Queensland mastheads. Star News said: 'In the future, the sky's the limit for Today Print, with plans to print other newspapers from throughout Queensland and pick up commercial work' (Pearce, 2021). The interviewees from *Gympie Today* said this was a source of comfort for them as it indicated the intention of the company to invest in the longevity of regional Queensland journalism.

News Corp's withdrawal from Queensland regional journalism was a surprise. For their part, News Corp management described the sale as part of the consolidation of exciting digital journalism opportunities for Queensland (Meade, 2020). It remains to be seen what these exciting opportunities might be, but it is clear Star News Group is in the ascendency in Queensland regional print journalism.

## 5 Same as It Ever Was?

Both Jarrett and Gorrie said they were proud that all the content in their respective publications was local. None of what was printed was sourced from the other *Today* publications or from news agencies such as Associated Press, Australian Associated Press or Reuters. Both said there was no budget to pay for news agency copy making the quest to produce truly local journalism driven by a mixture of pride and budget.

All the former News Corp journalists interviewed agreed that a frustrating aspect of working there was the publication of syndicated copy across different publications. Coverage of local events and issues dwindled as did the prospect of doing any substantive investigative journalism in the regions. Newspapers such as News Corp's *Gympie Times* and *Noosa News* became disconnected from the communities they once served. It was in this context that Star News Group entered the Queensland regional news market.

But then, in a move that seemed contrary to the continual downsizing of their Queensland operations, in May 2021 News Corp announced it intended to employ an additional 100 editorial staff (Ward, 2021) to service its digital news offering as well as partnering with Google in establishing a digital journalism school for their staff (Mediaweek, 2021).

Aside from news.com.au, News Corp's Australian national news and journalism web aggregator, Queensland readers could access local, state, national and international journalism through the paywall-protected, the *Courier Mail* website. The conservative *Courier Mail* is the last remaining daily metro newspaper in Brisbane, the Queensland state capital. Three months later, in July, News Corp announced it would be broadcasting *Sky News* across regional Queensland for free. A month later, the conservative broadcaster was served a 1-week ban from YouTube after the tech giant had deemed *Sky* guilty of disseminating false and misleading information on COVID-19 on the platform (Meade, 2021). YouTube also took the opportunity to purge their platform of *Sky News* content they found violating their policies. At the time of writing it was not clear whether the pivot away from traditional community-based newspapers to paywall-protected journalism and an increased conservative-enhanced broadcast presence was a deliberate shift in platform or the last gasp of a failing giant.

Despite a clear desire on the part of Star to expand in Queensland, its operations were not without problems. In the interviews, it was clear that those working under the Star umbrella in Queensland were not completely satisfied with their work conditions—stressing they were doing a lot more with a lot less. Both *Gympie Today* and *Noosa Today* publish weekly. Even with seven-day publishing intervals, the interviewees said they were all overworked. As noted earlier, this was especially true for the women journalists. Maccoll and Jones were both on part-time contracts, but both said a normal working week usually saw them working more than twice their contracted hours. Despite working long and unpaid hours, both said there was always more work to be done and the pressure to do so was immense. The same time pressures were true for the young women occupying the entry-level positions at each publication. Although both Astley and Callan were full-time employees, both worked well beyond their paid hours.

## 6 Subsistence Journalism

Thus, a model of 'subsistence journalism' emerges predicated on exploitation that allows the owners to extract as much as possible from those beneath them. This is neither strange nor unexpected, but it does reveal an uncomfortable truth inherent in the political economy of local journalism—it is a business dependent on profits for survival.

As already noted, some of the participants felt overworked and stressed. They also said there was little to no prospect for professional development or career advancement. Again, this was especially true for the women occupying entry-level positions. Along with concerns about their longevity as journalists, both Astley and Callan were worried about becoming stagnant as journalists before their careers had begun. Astley said although she was able to draw on the experience of Jones and Gorrie, there was no formal professional development available to her. As already noted, the situation was worse for Callan, and she was doubtful there was a future for her journalism. In terms of professional development, the situation was not any better for Jones or Maccoll although they had the benefit years of experience before their current roles. For Gorrie and Jarrett, their careers were nearing their end making thoughts of career development redundant. This lack of investment in staff training in combination with long working hours made it difficult for the interviewees to imagine a change or improvement in working conditions and job expectations beyond considerations around getting the next edition published. To this end, what the two editorial teams were doing was subsistence in that their efforts were geared towards survival rather than growing, evolving and innovating within their respective spaces. According to the participants, this was reflected in the kinds of stories they were able to cover.

Jarrett said there was no end of interesting stories for him and his colleagues to cover but they lacked the time and resources to attend. Jarrett said:

> Just down there we have the Noosa River. It's beautiful but I would never swim in it or eat fish out of it. It's dirty. It's very polluted. I know why. Everyone around here knows why. I'd like to investigate but I don't have the time. There is never enough time for that kind of journalism.

He said there were many other stories in the Noosa community that demanded attention but would not be covered. These included stories about the local drug epidemic and the associated trade, about local government authorities ignoring illegal and troublesome developments, and about corruption in sports clubs. Jarret said that even if the staff were available to fulfil an investigative or public interest role at *Noosa Today*, there were no resources allocated to fight any legal challenges that could follow the publication of such stories. An uncomfortable truth emerges here. There is a clear willingness on the part of these journalists to investigate and tell the stories that sustain the democratic health of their community, but the means to tell such stories is not available. In the case of pollution and dumping in the Noosa River, this story goes to heart of sustaining local ecology and the people who rely on that waterway, yet this story will not read in *Noosa Today*. Also, this story will not be run out and fear of retribution and the effects such retribution could have on the sustainability of the paper and the people who work there.

This raises important questions about the role of journalism in the regional Queensland and beyond. Does the mere presence of a local masthead guarantee local powerholders will face the kinds of scrutiny that sustain robust democratic communities? The preceding example suggests the simple answer is 'no'.

No legal budget did not stop Gorrie from taking a more cavalier attitude to the prospects of a defamation suit. In Australia, defamation is a civil matter, and as Murray et al. (2021) show, it can be used as a weapon by the rich, power and elite to shut down or 'kill' a story. Gorrie was aware some of his work could unset people. Despite there being no budget to appraise a story for the chances of being defamatory before publication (the jargon term for this is 'pre pub') nor necessary legal provisions provided by Star, Gorrie had no plans to stop publishing stories in his community's public interest.

Under News Corp, access to legal advice pre-publication was available. From their base in Surry Hills on the edge of Sydney's central business district, News Corp retains a department of in-house lawyers dedicated to making sure any interactions with the law work out in the favour of News Corp, their journalists and their editorial staff. Star News Group does not have the capital to provide such services.

Lack of legal advice may not have stopped Gorrie from covering potentially legally fraught stories, but lack of staff did. This was common across both *Gympie Today* and *Noosa Today*. Publishing once a week meant that many stories from the respective communities went untold. This frustration was compounded by the perception on the part of all the interviewees that their journalism was profitable, yet these profits were not being returned *Gympie Today, Noosa Today*, nor the communities they served. Despite both editorial teams being overworked to the point of burnout, none of those interviewed had faith their working conditions would improve. One interviewee said:

> We all know we make plenty of money for Star. Where does it go? From what I can see it goes to propping up less profitable newspapers.

Herein lies the key issue of corporate ownership of regional news media. The popularity and success of a local or regional news publication do not always mean more money will be invested to make the publication bigger and better.

It is a cruel irony that under Star ownership, journalists working on *Today* mastheads can set the agenda of their publications but lack the resources to do so. According to Gorrie and Jones, this is in contrast to their former lives with News Corp where who and what could be covered or reported was carefully controlled by more senior journalists and executives in Brisbane, Sydney, and beyond. Under current conditions, the journalists interviewed are freer than they were, but they are constrained by time and money and pressures and publishing a fresh edition every week. There is a willingness and a desire to chase the kinds of stories that enrich democracy, but the decoupling of democracy and the fourth estate Deuze and Witschge (2020) describe happens in regional Queensland because the resources needed to tell these stories are not available.

## 7 Conclusion

Star News Group, it appears, provides just enough staff and resources to get a paper out every week. That alone sees staff working above and beyond their contracted hours to breaking point. There is no provision for staff training, more staff or legal advice. While Star provides editorial, publishing and distribution infrastructure, and each of the publications is able to set the tone of their respective editorial agendas, there are few possibilities for those working in these newsrooms to do more or better than they are now. Contrary to the efforts of editorial staff, Star's lack of further investment results in local journalism that is predictable and shallow.

It was clear in the interviews with the journalists at *Noosa Today* and *Gympie Today* was the desire to function in the mould Papachrissi's (2021) 'democracy's detectives' is alive and well. Jarrett and Maccoll's willingness to express their frustrations on the stories they felt they could not tell revealed a lot about the kinds of journalism they wanted to do. For Gorrie, the frustration had turned to action. In order to reveal local and state government duplicity in dealing with the bushfires that had devastated Fraser Island, he had used his own funds to charter a helicopter to break a state imposed 'no fly zone' over the island. In doing so, Gorrie revealed the true extent of the fires and held his local and state authorities to account over hiding the scope of the fires. This issue here was Gorrie was acting on his own initiative and spending his own money to tell a story of national significance. While happy to publish his story, his employer would not fund such journalism.

Subsistence journalism leads to a situation where a local news publication is present but unable or unwilling to take on stories that challenge authority and support democracy. Under these conditions the illusion of journalism exists. However, the reality is, this form of journalism's sole goal is survival. This means filling the news hole with very limited resources with whatever it takes to fill a paper every week. It is early days for *Noosa Today* and *Gympie Today*. At the time of the interviews, both were less than a year old. Months shy of their first birthdays, it was impossible to know what the future held for these newspapers, and the journalists who tirelessly who under these mastheads for the communities they love. What is clear, is if the journalists at *Noosa Today* and *Gympie Today* are not properly supported in telling the stories they yearn to tell in the interests of sustaining healthy democratic communities, a new news desert will emerge where the presence of local journalism does not result in journalism that matters.

## References


Abernathy, P. M. (2016). *The rise of a new media baron and the emerging threat of news deserts*. Center for Innovation and Sustainability in Local Media, Chapel Hill, University of North Carolina Press.

Abernathy, P. M. (2020). *News deserts and ghost newspapers: Will local news survive?* University of North Carolina Press.

Australian Competition and Consumer Commission. (2016). ACCC will not oppose News Corporation's proposed acquisition of APN's Australian regional media division – ARM, Australian Competition and Consumer Commission, Government of Australia. Accessed December 7, 2016, from https://www.accc.gov.au/media-release/accc-will-not-oppose-news-corporation%E2%80%99s-proposed-acquisition-of-apn%E2%80%99s-australian-regional-media-division-%E2%80%93-arm

Butler, B. (2021). News Corp recovers from horror year with $445m profit. *The Guardian Australia*. Accessed November 2, 2021, from https://www.theguardian.com/media/2021/aug/06/news-corp-recovers-from-horror-year-with-445m-profit

Deuze, M., & Witschge, T. (2020). *Beyond journalism*. Wiley.

Dodd, A., & Ricketson, M. (Eds.). (2021). *Upheaval: Disrupted lives in journalism*. NewSouth Publishing.

Downman, S., & Murray, R. (2017). *Hyperlocal journalism and digital disruptions: The journalism change agents in Australia and New Zealand*. Routledge.

Fisher, C., Park, S., Howden, S., Lee, J. Y., & McGuinness, K. (2020). *Australian regional journalists: What they need and how they see the future*. Canberra University.

Hanusch, F. (2012). Broadening the focus: The case for lifestyle journalism as a field of scholarly inquiry. *Journalism Practice, 6*(1), 2–11.

Hanusch, F. (Ed.). (2014). *Lifestyle journalism*. Routledge.

Hess, K., & Waller, L. (2016). *Local journalism in a digital world*. Macmillan International Higher Education.

Hess, K., & Waller, L. J. (2021). Local newspapers and coronavirus: Conceptualising connections, comparisons and cures. *Media International Australia, 178*(1), 21–23.

Kelly, V. (2020). News Corp closes over 100 print papers, with 14 titles to disappear completely. *Mumbrella*. Accessed May 28, 2020, from https://mumbrella.com.au/news-corp-closes-over-100-print-papers-with-14-titles-to-disappear-completely-629436

Kvale, S. (2012). *Doing interviews*. Sage.

Lamb, J. (2021). Noosa the most expensive housing market in Queensland. *ABC News*. Accessed March 24, 2021, from https://www.abc.net.au/news/2021-03-24/inside-he-most-expensive-property-market-in-queensland/100024004

Lidberg, J. (2019). The distortion of the Australian public sphere: Media ownership concentration in Australia. *AQ-Australian Quarterly, 90*(1), 12–20.

Madden, R. (2017). *Being ethnographic: A guide to the theory and practice of ethnography*. Sage.

Meade, A. (2020). News Corp announces end of more than 100 Australian print newspapers in huge shift to digital. *Guardian Australia*. Accessed May 28, 2020, from https://www.theguardian.com/media/2020/may/28/news-corp-announces-end-of-nearly-100-australian-print-newspapers-in-huge-shift-to-digital

Meade, A. (2021). Sky News Australia banned from YouTube for seven days over Covid Misinformation. *Guardian Australia*. Accessed August 1, 2021, from https://www.theguardian.com/media/2021/aug/01/sky-news-australia-banned-from-youtube-for-seven-days-over-covid-misinformation

*Mediaweek*. (2021). News Corp and Google to launch the digital news academy program, *Mediaweek*. Accessed August 2, 2021, from https://www.mediaweek.com.au/news-corp-and-google-to-launch-the-digital-news-academy-program/?fbclid=IwAR1QTIvQBDLw-td5diupNT0AAjvqeWUgOzu4FqZhwmptjfqVlwiZfjrDe8Y

Mill, J. S. (1989). *On liberty*. Cambridge University Press.

Milton, J. (2014). *Areopagitica and other writings*. Penguin.

Murray, R., Ananian-Welsh, R., & Greste, P. (2021). Journalism on ice: National security laws and the chilling effect in Australian journalism. In T. Workneh & P. Haridakis (Eds.), *Counter-terrorism laws and freedom of expression: Global perspectives* (pp. 173–191). Lexington Books.

News Corp Australia. (2020). News Corp Australia announces portfolio changes, newscorpaustralia.com. Accessed May 27, 2020, from https://www.newscorpaustralia.com/news-corp-australia-announces-portfolio-changes/

North, L. (2009). *The gendered newsroom: How journalists experience the changing world of media*. Hampton Press.

Olsen, R. K., Pickard, V., & Westlund, O. (2020). Communal news work: Covid-19 calls for collective funding of journalism. *Digital Journalism, 8*(5), 673–680.

Örnebring, H. (2018). Journalists thinking about precarity: Making sense of the 'new normal'. *International Symposium on Online Journalism, 8*(1), 109–127.

Papacharissi, Z. (2021). *After democracy*. Yale University Press.

Park, S., Fisher, C., McGuinness, K., Lee, J. Y., & McCallum, K. (2021). *Digital news report: Australia 2021*. Canberra University Press.

Pearce, M. (2021). Queensland print revival, *Noosa Today*. Accessed June 1, 2021, from https://noosatoday.com.au/news/01-06-2021/queensland-print-revival/?fbclid=IwAR2dmXq5E0GCW0v7LvINbPsusqcvQffGcIdi0deibUavJA9Vofe2Y4vO5Dk

Picard, R. (2014). Twilight or new dawn of journalism? Evidence from the changing news ecosystem. *Journalism Practice, 8*(5), 488–498.

Pickard, V. (2019). *Democracy without journalism*. Oxford University Press.

Public Interest Journalism Initiative. (2021). The Australian newsroom mapping project, *Public Interest Journalism Initiative*. Accessed August 1, 2021, from https://anmp.piji.com.au/

Schulz, A. (2021). Local news unbundled: where audience value still lies. *Reuters Institute*. Accessed November 1, 2021, from https://reutersinstitute.politics.ox.ac.uk/digital-news-report/2021/local-news-unbundled-where-audience-value-still-lies

Star New Group. (2021). *About Star News Group Ltd*. Accessed August 13, 2021, from https://www.starnewsgroup.com.au/about/

Villi, M., & Picard, R. (2019). Transformation and innovation of media business models. In M. Deuze & M. Prenger (Eds.), *Making media: Production, practices, and professions* (pp. 121–132). Amsterdam University Press. https://doi.org/10.1515/9789048540150-009

Wahl-Jorgensen, K. (2019). The challenge of local news provision. *Journalism, 20*(1), 163–166.

Ward, M. (2021). News Corp Invests in regional news one year after move to digital only. *Australian Financial Review*. Accessed May 10, 2021, from https://www.afr.com/companies/media-and-marketing/news-corp-invests-in-regional-news-one-year-after-move-to-digital-only-20210507-p57pxg

Wilczek, B. (2019). Complexity, uncertainty and change in news organizations. *International Journal on Media Management, 21*(2), 88–129.

# Journalism and Ethics Amid the *Infodemic*

**Peter Greste**

On May 19, 2020, as the COVID-19 epidemic was spinning out of control around the world, the then US President Donald Trump took his seat at an event at the White House, and mentioned to reporters he would been taking the anti-malaria drug hydroxychloroquine daily to prevent coronavirus infection.

"I happen to be taking it," Trump said.[1] "A lot of good things have come out. You'd be surprised at how many people are taking it, especially the front-line workers. Before you catch it. The front-line workers; many, many are taking it".

In March, two months earlier, the drug emerged as a possible candidate for treating COVID-19 after a statistically tiny non-randomised study with 36 patients found "a rapid and effective speeding up of their healing process, and a sharp decrease in the amount of time they remained contagious".[2] The scientific community greeted the news with scepticism,[3] but the study triggered a wave of interest around the world[4] and prompted the president to push the drug in another news conference where he admitted he was acting on "gut instinct" before asking, "what have you got to lose?"[5]

Peter Greste is Professor of Journalism at Macquarie University.

[1] https://www.cnbc.com/2020/05/18/trump-says-he-takes-hydroxychloroquine-to-prevent-coronavirus-infection.html

[2] https://www.sciencedirect.com/science/article/pii/S0924857920300996?via%3Dihub

[3] https://theconversation.com/when-trump-pushed-hydroxychloroquine-to-treat-covid-19-hundreds-of-thousands-of-prescriptions-followed-despite-little-evidence-that-it-worked-140156

[4] https://www.ncbi.nlm.nih.gov/pmc/articles/PMC7685699/

[5] https://www.washingtonpost.com/politics/what-do-you-have-to-lose-inside-trumps-embrace-of-a-risky-drug-against-coronavirus/2020/04/06/0a744d7e-781f-11ea-a130-df573469f094_story.html

P. Greste (✉)
Macquarie University, Macquarie Park, NSW, Australia
e-mail: peter.greste@mq.edu.au



F. Weder et al. (eds.), *CSR Communication in the Media*, CSR, Sustainability, Ethics & Governance, https://doi.org/10.1007/978-3-031-18976-0_15

The news, and the president's remarks, triggered a rush in demand for hydroxychloroquine, with people around the world clearing the drug off pharmacy shelves and triggering a dangerous shortage for those who rely on it to treat the autoimmune disease, lupus.[6] Meanwhile, hydroxychloroquine remained unproven as a prophylactic, and with its serious side effects, including muscle weakness and heart arrhythmia,[7] many other researchers warned that taking it could be harmful.[8]

For journalists committed to reporting fairly, accurately, with balance, and in the public interest, the moment presented a particularly awkward challenge. By quoting President Trump, the journalists knew they were amplifying a dangerous myth. Every mention of the drug, even negatively, kept it at the front of people's minds, encouraging many to ask their doctors for prescriptions and in the process, unnecessarily exposing them to the risk of side effects. Yet as the most powerful person in the world, the president's remarks *always* count as "news", and a failure to report them would have exposed journalists to allegations of censoring the president for political reasons. Even placing the president's comments in context by emphasising the dangerous nature of his unproven claims, merely added fuel to the social media fire.[9] It was "fake news" at its most vexing.

But there was an added complication. Where once journalists were the main pipeline for "news", now they are merely one drop in the firehose of information streaming out of smart phones everywhere. Anyone who wanted to know the President's thoughts on the matter could bypass the reporters and their news organisations altogether and go straight to @realDonaldTrump on Twitter, or to one of the many of his equally unqualified amplifiers.[10] And for the millions who got their news principally from social media, many would have avoided seeing a story on the subject written by *any* credible journalist.

## 1 The Death of the *Constitution of Knowledge*?

The tale emphasises the radical challenge the digital revolution poses to what the American author Jonathan Rauch (2021) calls the "Constitution of Knowledge". Rauch argues that that the "reality-based community [is] the social network which adheres to liberal science's rules and norms ... objectivity, factuality, rationality: they live not just within individuals' minds and practices but on the network". The

---

[6] https://www.theguardian.com/world/2020/mar/27/vital-drug-people-lupus-coronavirus-covid-19-link-hydroxychloroquine

[7] https://www.mayoclinic.org/drugs-supplements/hydroxychloroquine-oral-route/side-effects/drg-20064216

[8] https://www.healthline.com/health-news/trump-is-taking-hydroxychloroquine-why-experts-think-this-is-a-bad-idea

[9] https://www.cjr.org/the_media_today/big_tech_hearing_coronavirus_misinformation.php

[10] https://journals.sagepub.com/doi/10.1177/20563051211024977

community he speaks of is the collective of liberal institutions including not only the hard sciences but also academia, journalism, government, and the law. Together, they form what he calls a "marketplace of persuasion", driven by the classical search for objective truth, in a way that underpins modern democracy.

Rauch argues that the Internet, and social media companies in particular, have created conditions that are unravelling the *Constitution of Knowledge,* and in the process triggering a wave of angst in liberal democracies everywhere. He describes it as an "epistemological crisis", and while he acknowledges that it has many causes including the anarchy of the Internet and left-wing cancel culture, he places the weight of responsibility on President Trump's shoulders. To Rauch, Trump and his populist associates consistently assaulted the truth, "because their goal was to denude the public's capacity to make any distinctions". The President confirmed as much when he told CBS 60 Minutes reporter Lesley Stahl, that he constantly attacked the press to "demean you all, so when you write negative stories about me, no one will believe you".[11]

It is perhaps a bit unfair to blame one politician, even if, for four years, he was the most powerful figure on earth. The crisis Rauch speaks of could not have happened without the capacity to use the internet to bypass that Constitution of Knowledge; to speak directly to a community without having to bother with the ethical filters that (however imperfectly) regulated the flow of information to create the foundation for public debate. As Republican Senator Ben Sasse once told CNN, "We have a risk of getting to a place where we don't have shared public facts. A republic will not work if we don't have shared facts".[12]

In the pre-Internet days, much of those shared facts flowed through journalists who controlled and directed information to readers and audiences. In his book, "News 2.0; can journalism survive the Internet?", academic Martin Hurst wrote, "In truth, up until a few years ago, whether we liked or trusted them or not we relied on journalists for all our news information. We didn't even have to think about it. Our news was like life insurance and banking: solid, reliable, and not too flashy".

The American journalism professor David Manning White was the first to use *Gatekeeper theory* to model the way journalists decided what stories and information deserved to be included in their output, and what ones should go in the bin. In his original 1950 study, White asked a regional editor (who he cheekily nicknamed, "Mr Gates") to keep all the copy that came into his office from three wire services in one week, and explain why he decided to reject certain stories. White found that Mr. Gates made news judgements based partly on his subjective experience as an editor, but also on his attitudes and expectations. Some stories were rejected because they duplicated other reports, while others were rejected because Mr. Gates decided they were "not newsworthy" according to his own assessment. He admitted to preferring political news to other types, tried to avoid sensationalism, and did not like suicide stories. Although White's study has been widely criticised for ignoring

[11] https://globalnews.ca/news/7409696/trump-60-minutes-walkout/

[12] https://transcripts.cnn.com/show/sotu/date/2017-07-02/segment/01

other pressures and influences on the decision-making process, it remains one of the seminal contributions to communications theory.

While White did not explicitly include it in his study, there is little doubt that Mr. Gates would also have been making decisions according to journalistic ethics such as accuracy, balance, fairness, and so on. The editor might have been doing so out of high-minded principle, but he would also be doing it to protect his own reputation and that of his newspaper. Anything patently wrong, or clearly biased would have been weeded out. No journalist could expect to remain employed for long if they had deviated from those professional norms and standards. Together, they acted as a kind of self-regulating if flawed mechanism for protecting the integrity of the information pipeline from those institutions that formed the Constitution of Knowledge, to the public, via the news media.

The concept of the gate conferred enormous power and influence on journalists, giving them de facto authority to decide who and what gets communicated to audiences. It is a form of control over "the public's eyeball", directing attention to key issues of the day, both feeding and provoking public and political debate. But that power was not absolute, and it is important to not over-state the influence of any particular individual (though editors in charge of the mass-circulation dailies clearly wielded significant power). Rauch's liberal knowledge community also generally formed a consensus about the key topics for public debate, maintaining a consistency of both the subjects people talked about, and the basic facts that informed their conversations. The consistency of topics covered on the front pages of most dailies and the headlines on most mainstream TV news services suggests that the gatekeeping system worked collectively to set the agenda for what people discussed over the dinner tables.

In their book, *Gatekeeping Theory,* academics Shoemaker and Vos (2009) succinctly described it as the "process of culling and crafting countless bits of information into the limited number of messages that reach people every day, and it is the centre of the media's role in modern public life. [...] This process determines not only which information is selected, but also what the content and nature of the messages, such as news, will be".

In the early days of the Internet's inception, anyone who suggested this was a good thing was condemned by free speech advocates and cyber utopians as an apologist for powerful media elites. Those idealists argued the Internet was a democratising force, opening access to both the sources of information and the means of publication to anyone with a keyboard and an Internet connection. The Internet would, they said, challenge the corporations that controlled the flow information for their own interests. But curiously, Shoemaker and Vos wrote their description in 2009, well after the digital revolution had already smashed down the traditional gates that journalists were supposed to be keeping. Three years *earlier*, another academic, Jane Singer (2006), argued that "the power of gatekeepers seems to diminish in a modern information society. The Internet defies the whole notion of a 'gate' and challenges the idea that journalists (or anyone else) can or should limit what passes through it".

In mounting her argument, Singer assumes gatekeepers must be human, but even in a completely digital world, the gates and their keepers remain. Information does not simply pour out of our online devices, unmediated and disorganised. Every search engine and social media company employs algorithms to decide what bits of information to send us and in what order, depending on our browsing and search histories. For example, Google uses more than 200 "ranking factors" to decide how to sort through the billions of pages online and deliver what it deems as the most relevant links to users.[13] Journalists and their news values are no longer in charge. Gone are the ethics behind editorial judgements, along with the principles of accuracy, balance, or fairness, (the Internet is, after all, supposed to be "value-neutral"). Instead, the parameters are set by engineers whose sole purpose is to create a stream that holds our eyeballs. So rather than producing a diet of information curated to provide a mix of content across a range of issues and direct our attention to areas of common debate in the way that classical newspapers and their websites once did, social media algorithms deliver a pipeline of individually tailored narrow-cast titbits of information (I cannot quite bring myself to call those titbits "news".)

In 2015, the University of Calgary researcher Emily Laidlaw coined the term, *internet information gatekeepers* (IIGs), to describe the way that a digital gatekeeper "facilitates or hinders deliberation and participation in the forms of meaning-making in democratic culture".[14] In her paper, she points out:

> Every time we use the internet, we engage with IIGs. To find information, we use search engines. To access the internet, we need to use internet service providers. To be able to participate on message boards or social networking sites, we go through a host. The role of such regulators has not yet been settled, and, as of yet, *they do not have any democratic or public interest mandate that assures the internet's democratic potential is being facilitated.* If the internet is a democratising force, we inevitably at present must rely on these IIGs for the realisation of this aspect of its capacity. It is argued ... that the corporate social responsibility frameworks that currently govern the activities of IIGs are insufficient to meet their human rights obligations and that, without intervention, the continuation of their work in its current mode will hamper the ability of the internet to work as a tool of democracy. (Emphasis added.)

In other words, because those gatekeepers are untethered from the moral and ethical guardrails that kept the classical Constitution of Knowledge honest, they are either directly or indirectly undermining democratic debate. We are not only losing much of the common basis for public discussion; as Rausch argues, in some cases, Internet technologies directly undermine it.

---

[13] For example, see https://backlinko.com/google-ranking-factors. For Google's own explanation of the ranking system, see https://www.google.com/search/howsearchworks/algorithms/

[14] https://www.cambridge.org/core/books/regulating-speech-in-cyberspace/internet-as-a-democratising-force/AEFBA8BCA0CD6CDB3E4AD61B706D1BCF/core-reader#

## 2 The *Infodemic*

The World Health Organisation (WHO) believes we have something of a dual-pandemic; a combination of the COVID-driven health crisis, and what the WHO has described as an *infodemic*: "too much information including false or misleading information in digital and physical environments during a disease outbreak".[15] The WHO says the *infodemic* causes confusion and risky behaviours that can harm health just as surely as the real-world virus. It leads to mistrust in health authorities and undermines the public health response, intensifying or lengthening outbreaks when people are unsure about what they need to do to protect their health and the health of those around them. With growing digitisation, information—and more critically, *misinformation*—spreads more rapidly, filling information voids with toxicity, amplifying harmful messages, and diluting or discrediting good information. That is exactly what Trump admitted doing in his remark to Lesley Stahl.

The same is true of the climate change debate. Deliberate Trumpian attempts to discredit and undermine legitimate science (by both the president and other denialists) has clouded both public debate and the political response. It has made it far harder for governments to take the necessary but difficult policy decisions that might slow or reverse climate change. Another team from the University of Exeter explored how climate change misinformation spreads online and, in their paper, they described finding certain political and religious organisations, foundations, and corporates that worked to cloud the debate by exploiting social media algorithms tailored to trigger some very human behaviours.[16] Sometimes it was both deliberate and cynical; at other times it was amplified by the design of the social media algorithms themselves.

> The spread of misinformation is intertwined with a number of online and offline social processes. One of these is 'homophily'—the tendency for people to form social connections with those who are similar to themselves, as captured by the common saying 'birds of a feather flock together'... This behaviour is encouraged by social media platforms in the way new connections are recommended. ... Another factor which can contribute to polarisation is the way online social networks promote content based on being engaging and aligned with your previous viewed material rather than on trustworthiness. This is known as "algorithmic bias" and amplifies the psychological finding that people tend to prefer to consume information that matches their belief systems—known as "confirmation bias".

It all leads to what the Exeter team colourfully described as a "paralysing fog of doubt".

Amid *that* world, suddenly old-fashioned journalism starts to look a lot more appealing, and not just to journalists. The public more broadly appears to be looking for routes out of the fog by following established news sources. In March 2020, at the height of Australia's first COVID lockdown, the ratings company Nielsen reported that the top ten news sites saw an average increase in readers of 57%

[15] https://www.who.int/health-topics/infodemic#tab=tab_1

[16] https://wires.onlinelibrary.wiley.com/doi/full/10.1002/wcc.665

compared with February, before the lockdown began.[17] And while before the pandemic the country's major news sites all routinely struggled to reach more than 10 million unique visits, during the lockdown seven of the top 10 were reporting over 11 million visitors; all recorded their highest unique audience figure since Nielsen started their digital ratings. In the United States, legacy news organisations saw an almost identical spike. The New York Times reported that overall visits to news sites rose by 57% in March 2020 over the same period a year earlier, and the number of minutes that readers spent on those sites surged 46%.[18] The outlets that showed the biggest gains were all the trusted legacy brands—*The Atlantic*, *Business Insider*, *The New York Times*, *The Los Angeles Times*, *The Wall Street Journal*—all of which doubled or nearly doubled the number of visits. (In a curious irony, the number of unique visitors to Vanity Fair soared 44% in one week, thanks in part to a tweet by President Trump, attacking the magazine's coronavirus coverage.)

John Micklethwait, the editor in chief of *Bloomberg News* told the *New York Times*, "People are really desperate for factual, good information and analysis. My guess is a lot of the quality end of journalism has done well, because suddenly it matters enormously what the facts are".

But consuming the news is not the same as *trusting* the news. Just because people are rushing to the legacy news services does not mean they have any greater confidence in what they are reading. In February 2021, after most of the world had been through the first wave of the virus and we saw the related surge in visits to news services, the statistical website Satista asked consumers how much they trusted their news media. Of the 40 countries included in their survey, only 13 scored 50% or higher of those who said they trusted the news "most of the time".[19] Finland led the pack with 65%, followed by Portugal and Kenya (both 61%), and Denmark and Netherlands (both 59%). Australia rated a paltry 43% and the United Kingdom just 36%. The United States sat at the very bottom of the survey with only 29% of people trusting the news media most of the time.

The annual Edelman Trust Barometer found a similar trend. In the introduction to its 2021 report Edelman said, "After a year of unprecedented disaster and turbulence—the Covid-19 pandemic and economic crisis, the global outcry over systemic racism and political instability—the 2021 Edelman Trust Barometer reveals an epidemic of misinformation and widespread mistrust of societal institutions and leaders around the world. Adding to this is a failing trust ecosystem unable to confront the rampant infodemic, leaving the four institutions—business, government, NGOs and media—in an environment of information bankruptcy and a mandate to rebuild trust and chart a new path forward".[20]

---

[17] https://www.nielsen.com/au/en/press-releases/2020/australians-turn-to-digital-news-to-stay-informed/

[18] https://www.nytimes.com/2020/03/20/business/coronavirus-news-sites.html

[19] https://www.statista.com/statistics/308468/importance-brand-journalist-creating-trust-news/

[20] https://www.edelman.com/trust/2021-trust-barometer

## 3 The Death of *Objectivity*

One of journalism's touchstones is "objectivity"—the idea that it is both possible and desirable for journalists to make no judgement about the stories they are covering. Instead, they should merely observe, record, and report the facts and events as they unfold. In the process, journalists are supposed to suspend any opinion or subjective choices of their own, behaving as the "eye of God", documenting "the truth", without interference. It is a nice idea, but hopelessly out of touch with the realities of how journalists work. (The great American writer, Hunter S Thompson (1973) once wrote, "So much for Objective Journalism. Don't bother to look for it here—not under any by-line of mine; or anyone else I can think of. With the possible exception of things like box scores, race results, and stock market tabulations, there is no such thing as Objective Journalism. The phrase itself is a pompous contradiction in terms.")

As Thompson hints, journalists are *always* making subjective judgements. It is impossible not to. From choices about the questions a journalist asks to the questions they do not, from the information included and the details junked, to the order that information is presented, subjective choices are utterly unavoidable at every stage of the process. That extends to people who journalists consider to be credible sources worth talking to for a story, over the unqualified or the cranks who are rejected. In a pre-Internet world, anyone who did not get over a journalists' credibility bar could still tell their friends and print off leaflets to hand out on a street corner, but they would otherwise exist on the fringes beyond the sight of the vast majority. Now though, those like Australian celebrity chef Pete Evans have been able to build a huge following on social media, peddling dangerous anti-vaccine theories and "cures" for COVID-19. (Evans was heavily fined for doing so.[21]) Because he is a minor celebrity, Evans is still able to reach millions of followers and argue that the media is biased because it refuses speak to him or take his ideas seriously. Thus, that *fog of doubt* becomes even thicker.

That puts news organisations back in the position discussed at the beginning of this chapter: do they cover the climate change deniers, the anti-vaxers, or the 9/11 conspiracy theorists in the name of "balance" and "objectivity" and risk giving credibility to crazy and even dangerous ideas? Reporting Donald Trump's lies is one thing, but what about other noisy voices? Is it right for journalists to make explicit editorial decisions to ignore them? In the climate change debate, well before the pandemic added another layer of confusion, several leading news organisations decided there could be no false equivalence between climate science and its sceptics. After years of internal arguments, the BBC's Director of News and Current Affairs Fran Unsworth in 2018 circulated a memo acknowledging that the organisation gets

[21] https://www.theguardian.com/australia-news/2021/may/25/pete-evans-fined-80000-by-health-department-for-alleged-unlawful-spruiking-of-devices-and-medicines

coverage of climate change "wrong too often" and told staff: "You do not need a 'denier' to balance the debate".[22]

"Climate change has been a difficult subject for the BBC", Unsworth said in her memo. "We get coverage of it wrong too often". She then wrote: "Manmade climate change exists: If the science proves it we should report it". She went on to warn against what she called "false balance":

> To achieve impartiality, you do not need to include outright deniers of climate change in BBC coverage, in the same way you would not have someone denying that Manchester United won 2-0 last Saturday. The referee has spoken.

.

In September 2019, the editor of the Australian edition of The Conversation, Misha Ketchell came to the same conclusion. In a piece published on its website, Ketchell bluntly declared, "Climate change deniers, and those shamelessly peddling pseudoscience and misinformation, are perpetuating ideas that will ultimately destroy the planet. As a publisher, giving them a voice on our site contributes to a stalled public discourse".[23] He then went on to declare a zero-tolerance approach to climate change deniers and sceptics: "Not only will we be removing their comments, we'll be locking their accounts".

The *Guardian Australia*, and the Australian Broadcasting Corporation have both adopted similar, if rather less vehement positions. The ABC's editorial policy declares that reporting with impartiality means, "a balance that follows the weight of evidence".[24] It is a policy that deftly slaps down anyone who believes balance means giving equal weight to all opinions.

Ultimately, this is good journalism. It is also good for public debate and national politics, but it is not always commercially successful. Note that all those on the list are either publicly funded or in the case of both *The Guardian* and *The Conversation*, supported by not-for-profit foundations. The legacy news organisations that are run as profit-making businesses all have very similar codes of conduct, but in a piece for The Conversation, the University of Melbourne research fellow Dennis Muller found that none of the big Australian media businesses explicitly tackled the climate change conundrum as directly as the ABC, the BBC or *The Guardian*.[25] Their codes give them enough wiggle room to drift into politically dangerous but commercially successful territory. In his piece, Muller said, "From this perspective, media acquiescence in climate change denial, failure to follow the weight of evidence, or continued adherence to an out-of-date standard of impartiality looks like culpable irresponsibility".

---

[22] https://www.theguardian.com/environment/2018/sep/07/bbc-we-get-climate-change-coverage-wrong-too-often

[23] https://theconversation.com/climate-change-deniers-are-dangerous-they-dont-deserve-a-place-on-our-site-123164

[24] https://edpols.abc.net.au/policies/

[25] https://theconversation.com/media-impartiality-on-climate-change-is-ethically-misguided-and-downright-dangerous-130778

Dennis Muller is right. Journalism must be responsible and adhere to its core principles, not despite but *because of* the urgent crisis of information disorder. In the debate around sustainability, for example much has been written about the way journalism communicates the arguments and the need to personally connect audiences and readers with the issues, and I do not intend to explore that further here. But the central point remains—to be effective in driving well-informed public debate and meaningful political action, journalism should not be seen as advocating a particular position. Rather, it should maintain its traditional values of following the evidence and critically evaluating those arguments by holding to scrutiny people involved in the wider debate. What matters is that the journalism is neutral on the question of policy, while being rigorous on evidence, without falling into the trap of false equivalence. There is no ethical justification for giving a climate change denier equal prominence to a climate scientist, just as no credible news organisation would balance a NASA physicist with a flat-earther.

But what happens when that journalistic ideal bumps up against the *other* conjunction of "journalism" and "sustainability"—when journalism itself, and the values Muller advocates, risk disappearing in the face of commercial reality? Again, a wide discussion about news business models is beyond the scope of this book, but a few points are worth making.

In the early days of "free news" online, journalists relied on clicks to generate an income. Readers could open any story they liked without having to pay, but each click meant a few cents would flow in to the newsroom from advertisers displaying their products alongside the stories. That tended to reward the sensational journalism over the sober, speed over accuracy, rumour over facts—the kind of thing that would motivate someone to open a story regardless of its integrity. Now, news companies have moved to subscription-based models, putting their content behind paywalls, and relieving journalists from having to produce "click-bait". But in a conversation for the Judith Neilson Institute's podcast, "Journo", the Dean of the Colombia School of Journalism Steve Coll, said the shift to subscription-based business models still required journalists to convince readers their news is worth paying for.

"To get that level of engagement, you have to get inside people's identity, inside their passions... When you see the pretence of a neutral news voice giving way to more explicitly partisan voices, whether in print, digital or cable television, you're seeing the pull of new business models that are focussed on engagement and subscription", he said. "It's very hard for newsrooms to develop models that will succeed economically without paying any attention to this kind of tribalism".[26]

The point relevant to this book is that if journalism is to play its proper role in the public conversation around sustainability (and any other debate for that matter), it must find a way of financing itself, that keeps it insulated from both commercial pressure and political interests. That would give it the space to adopt strategies that might recover some degree of integrity and public confidence in its output and thus support democratic politics. Journalists, news media owners, readers, and

[26] https://jninstitute.org/education/journo-podcast/

governments must now ask a fundamental question: should news be a business, *or* a public service? It seems unlikely it can be both.

If we want to preserve the *constitution of knowledge* that Rauch articulated; if we want to maintain a well-informed public debate; if we still believe in a fourth estate role for the media, it seems inevitable that we will have to re-engineer news in a way that encourages and rewards high quality, ethical journalism, even (especially) if it is not particularly popular or commercially valuable. The industry cannot do that without the kind of widespread support that translates to political action. That support will only come if journalists raise their ethical standards in the first place, and win back at least some of the public trust that has eroded over the years.

In Reporters Without Borders' 2021 World Press Freedom Index, the Secretary General Christophe Deloire said, "Journalism is the best vaccine against disinformation. Unfortunately, its production and distribution are too often blocked by political, economic, technological and, sometimes, even cultural factors. In response to the virality of disinformation across borders, on digital platforms and via social media, journalism provides the most effective means of ensuring that public debate is based on a diverse range of established facts".[27]

The onus is as much on journalists to defend their professional integrity, as it is on regulators to defend their role.

## References

Rauch, J. (2021). *The constitution of knowledge: A defense of truth.* Brookings Institution Press.

Shoemaker, P. J., & Vos, T. P. (2009). *Gatekeeping theory*. Routledge.

Singer, J. B. (2006). Stepping back from the gate: Online newspaper editors and the co-production of content in campaign 2004. *Journalism and Mass Communication Quarterly., 83*(2), 265–280.

Thompson, H. S. (1973). *Fear and loathing: On the campaign trail '72'*. Straight Arrow Books.

[27] https://rsf.org/en/2021-world-press-freedom-index-journalism-vaccine-against-disinformation-blocked-more-130-countries

# Beyond State and Market Failure

## Corporate Social Responsibility as an Elementary Component of the Mission and Public Value Management of Public Service Broadcasters

**Wolfgang Reising**

## 1 Changed Framework Conditions

As a result of digitalization and networking and the accompanying social change, the media landscape is undergoing a sometimes crisis-ridden upheaval, which is both an opportunity and a challenge for traditional media companies and for public service broadcasting. The structures of supply and competition that have been established over decades are being subjected to an almost "tectonic shift."

This development also affects the established construct of the dual broadcasting system, which defines broadcasting in Germany as an area of tension between cultural assets and economic assets. On the one hand, this is determined by a still national media and broadcasting policy, but on the other hand, it is increasingly determined by an oligopoly of private and globally operating digital corporations. This is accompanied by changed forms and structures of social communication and its mediation.

If the public broadcasters in Germany want to continue to play a central role in the individual and public opinion- and will-forming process in a democratic state with a pluralistic society in times of a convergent media world with a changed intensity of competition and modes of reception, and if they want to continue to guarantee their codified programming mandate in the future, they need to make adjustments in line with the target system and in a solution-oriented manner.

W. Reising (✉)
Leipzig, Saxony, Germany

 
F. Weder et al. (eds.), *CSR Communication in the Media*, CSR, Sustainability, Ethics & Governance, https://doi.org/10.1007/978-3-031-18976-0_16

## 2 New Demands on Governance

These framework conditions, which are largely predetermined for the broadcasters, require further improvement in entrepreneurial and goal-oriented management, both in terms of effectiveness at the level of the objective and efficiency at the level of the formal goal.[1] In the management of instability in a media system increasingly shaped by online media, which blurs the traditional boundaries between the classic media genres, between interpersonal and mass communication and between producer and recipient, the core issue is how quickly and successfully the public broadcasters can implement the digital transformation process and their associated social responsibility, even in a harsher environment, without giving up their identity and the dominance of meta-economic objectives derived from the organizational purpose.

*Public broadcasters therefore must realize a social responsibility in two respects*: On the one hand, as a public cultural and confidential asset in the broadcasting sector, to guarantee the assigned function and maxim as a basic provider, and on the other hand, as a part of infrastructure or economic policy and, to a growing extent, to document accountability for its corporate acceptance and responsibility.

In this context, management is confronted with strategic and operational questions that focus on the overriding purpose of public service broadcasting, the mission, and its serving function for the citizen and society as the central starting point. Integrated into the common good-oriented axiom and the approach of multidimensional *public value* (PV)[2] understood as a public management concept, which focuses on the program mandate in the case of public service media companies, other aspects, in particular those of corporate responsibility, which are summarized by the concept of *corporate social responsibility* (CSR), must also be taken into account in the core business when taking a holistic view of social responsibility in practice. Specifically, the respective fields of action must be defined according to strategic guidelines, objectives must be balanced without conflict and actively managed in a goal-oriented manner. The latter also includes the following topics and the facets of corporate governance and corporate citizenship, which must be appropriately integrated into contemporary accountability when establishing legitimacy in dealing with stakeholders. Integrated into a process from performance-oriented thinking to impact-based management,[3] public broadcasters must not only secure the success and quality of their program offerings as a permanent challenge in internal processes and document[4] them vis-à-vis supervisory bodies, media policymakers and, in particular, the public, but also actively take this change into account in internal and external reporting.

---

[1] On the definition of effectiveness and efficiency in public broadcasting: Keuper and Brösel (2005), p. 6ff.

[2] Meynhardt (2008), p. 73.

[3] Gläser et al. (2016), p. 277ff.

[4] Frey-Vor (2017), p. 106ff.

# 3 New Demands on Reporting

The aim is to integrate public value and corporate social responsibility with the guiding principle of the social transformation process of sustainability in reporting. In general, the public value management approach is seen in the public sector as a further development of the dysfunctional traditional public administration characterized by impersonality and as an alternative or corrective to the deficits of the New Public Management (NPM) approach, which was also courted by politicians in Germany in the 1990s and originated in the Anglo-Saxon world. This reduces the citizen to a customer, contrasts the public and private sectors in an inappropriate way and disproportionately neglects non-economic or—at that time—not (yet) measurable values of output or impact.[5] The PV approach is characterized in particular by the performance indicators of the outcome or impact as "primary services,"[6] which can only come into being through an active role and through perception, i.e., the subjective perception of the individual, and thus represents an interaction with the citizen or society. "Public value can be understood as the value contribution that a media company makes to society over and above the value contribution for its own company (the "corporate value")."[7]

In its current form, the CSR concept is a global phenomenon that has been able to assume growing importance in theory and practice over the last two decades, not least due to the shortcomings of the shareholder value approach. "The concept of CSR asks about the social role of the company: Which actions are not only legal, but also ethically legitimate according to the demands of (civil) society based on common values? How can companies make a contribution to society that does justice to their role as economic and social actors?"[8] CSR thus understands organizations as parts of society and as social actors that assume social responsibility in the sense of sustainable development and, on a meta-level, has also led to a remarkable renaissance of the common good in Germany with its characterization of a social market economy.[9] Just as the CSR concept has permeated business, politics, and society, public broadcasters have also been involved in this process since the mid-2000s. The concept of *sustainability*, which has increasingly developed as a central social model and social transformation concept in recent years, also has a considerable impact on the management of profit and non-profit organizations and is expressed through the implementation of the management concept of corporate social responsibility in many companies. It can be seen in everyday language that,

---

[5] Alford and Hughes (2008), p. 133.

[6] Gläser et al. (2016), p. 34.

[7] Gläser and Kühnle (2020), p. 415.

[8] Trommershausen (2011), p. 126.

[9] In 2004, the BBC was the first public broadcaster to publish its report ("Living Public Value"), followed by the Australian ABC ("Commitment to green future") in 2006 and the ZDF in 2010 with its first sustainability report, and before 2020, at the national level, the ARD as a media association of the state broadcasters also published its first sustainability report.

not least due to early publications by the Club of Rome, the supranational agreements adopted by the UN in the early 1990s or their implementation in EU law or national legislation, the concept of sustainability is increasingly gaining ground in both the profit and non-profit sectors and is pushing the technical term of CSR into the background in the corporate environment, insofar as it has really been applied on a broad scale outside of specialist committees and forums.[10]

In Germany, against the background of the regulatory requirements of the so-called Corporate Social Responsibility (CSR) Directive for reporting on non-financial aspects of a company's activities, the question arises as to whether the obligation[11] in Germany, which has so far applied to a narrow circle of large private companies, usually operating on the capital market, does not automatically also apply to large public companies, which are per se committed to the common good. Even if this is currently not (yet) the case in Germany, public pressure is growing on these organizations to at least voluntarily and of their own accord demonstrate in a qualified manner, within the framework of sustainability reporting obligations, whether these institutions really do act in a socially responsible manner according to CSR criteria.

Apart from the conceptual vagueness of the two management concepts of public value and CSR, they have in common that they describe a path "to be taken" due to their processual character. Incorporated into a corporate governance system that includes both legally standardized and voluntary commitments, it is also important to implement meaningful *accountability*. In essence, this means conclusively answering the question of accountability, which is a difficult undertaking[12] in broadcasting due to the multidimensionality and multitude of soft factors and can be well described with the aphorism "Not everything that counts can be counted, and not everything that can be counted counts."[13]

Taking up the above-mentioned specificity of the target system of public service broadcasters with the particularity of a twofold social obligation, it can be said in the context of the overarching vision of the overall social transformation process and in analogy to the relationship between the objective and the formal objective level, the relationship between the two concepts is basically as follows (cf. Fig. 1).

Based on the legally codified and original purpose of guaranteeing the meta-economic objective of the basic service, it is to be ensured that through the implementation of the program mandate as the brand core and the reception or usability of the program bouquet, both an individual consumer value and, on the collective level, a media citizen or public value, i.e., also a social added value, is generated. The latter is to be grasped in its entirety, i.e., it goes beyond the corporate value of public broadcasters, which are already committed to the common good by their very origin,

[10]Mohr and Schiller (2021), p. 2.

[11]German Commercial Code (HGB), in particular, §§ 289b para. 1, 315b para. 1, § 340a para. 1a, and 341a para. 1a.

[12]Crane et al. (2010), p. 65.

[13]Einstein (1946).

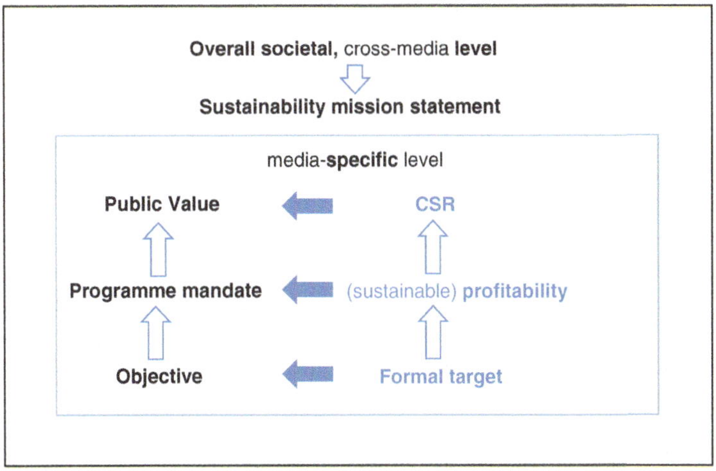


**Fig. 1** Relationships between sustainability, public value, and CSR concepts in public service broadcasting (Source: Own representation)

or the brand value of the organization, and is indispensable as its perceived value contribution to the citizen and society. Constitutive for the definition of public values are the three aspects of the journalistic value contribution (diversity), the social value contribution (identity), and the economic value contribution (value creation).[14]

Integrated into the context of the objective level as the overarching objective and the primarily socially oriented objectives (mission, vision, guiding principles), the voluntary commitment to implement the management concept of *CSR* is implemented at the formal objective level to promote sustainability in economic, ecological, and social terms. The formulation and anchoring of an individually tailored CSR strategy defines the scope and areas of social commitment and is at the same time an expression of the change in social responsibility, in which it is not only oriented toward the usual principles of good corporate governance in the private sector but also toward the Public Corporate Governance Code. CSR supports the objective of *sustainable economic efficiency* and can, among other things, flank the justification for the existence of public broadcasters and at the same time ensure their acceptance and credibility as social actors in society.

## 4 Reporting in the Context of Accountability

The reporting system of the broadcasting organizations, in particular, their annual reports, provided and still provides insufficiently qualified information on the efficient operationalization and the effective degree of fulfilment of the public

[14] Gläser et al. (2016), p. 43ff.

welfare-oriented objective in certain fields of action. It is, therefore, necessary to supplement the monitoring based on meaningful planning and the reporting system traditionally characterized by cost and performance accounting internally with elements which, in addition to the purely quantitative data of the input–output relationship from the micro level of the program as a basis to the meso level of the broadcaster, provide information about the performance impact goals and complex success mediation.

This is new territory both in theory and in practice, as the interweaving of relevant nonfinancial performance indicators makes accountability multidimensional and can draw on accountability conventions to a much lesser extent. This is because, up to the present, an asymmetry of information based on purely quantitative or monetary indicators has prevailed in reporting. But for integrated overall management, success in achieving self-imposed goals and media public value, it is no longer sufficient to efficiently manage resources and services, or the input–output ratios analyzed by KEF[15] in Germany. In view of the growing intensity of competition, public broadcasters must more than ever effectively fulfil their public task and objective of journalistically representing citizens and society in all their facets and reaching them through majority- and minority-oriented program offerings, yet non-profit organizations can also fail highly efficiently if they do not effectively manage their original organizational purpose and brand core.

Therefore, for holistic management that focuses on impact orientation, journalistically relevant guiding currencies such as the qualified presentation of reputation, credibility, trust, and the personal willingness to recommend for the citizen or society are voluntarily increasingly used in the accountability. Within the framework of this process, the control and information systems necessary for management have been and are being actively developed further in order to put the broadcasting organizations in a position to optimally fulfil the constitutionally standardized program mandate while observing the principles of sustainable economic efficiency and thriftiness and thus to present the system-relevant objective of basic provision and media public value as the new key currency of accountability in a meaningful way. In times of the internet, contemporary adaptations of the instruments are therefore not optional, but obligatory, if the broadcasters want to be able to give answers to their socio-political effectiveness, sustainability, and ultimately legitimacy in a (media-)politically influenced environment. The social value creation of the program bouquet and corporate responsibility are thus central components of the performance measurement and management information system, which is designed to be coherent with the target system and specific to the organization, and which, with regard to reporting, focuses, as outlined in simplified form in Fig. 2, on the significantly more complex determination of success on the service provision and

[15]The abbreviation KEF stands for "Kommission zur Ermittlung des Finanzbedarfs der öffentlich-rechtlichen Rundfunkanstalten" (Commission for the Determination of the Financial Requirements of Public Service Broadcasters) and is an independent commission commissioned by the 16 federal states to determine the medium-term financial requirements of Germany's public service broadcasters on the basis of a liquidity-oriented method.

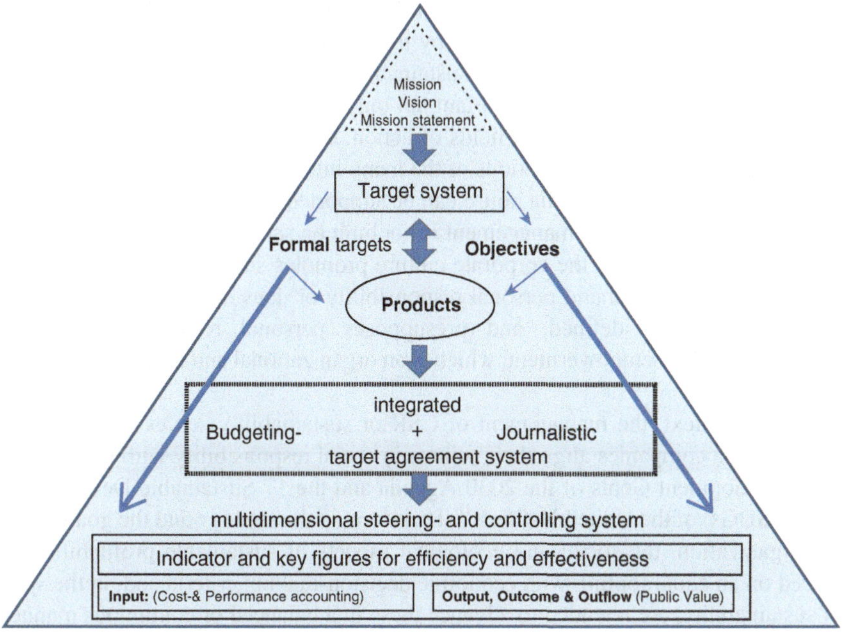


**Fig. 2** Transformation of the program mission along the service delivery and impact chain

performance effectiveness targets[16] at the product (micro), organizational (meso) and societal (macro) levels.[17]

## 5 Sustainability and CSR in Practical Implementation

Despite the increasing attractiveness of a broad conceptual understanding of the management concept of CSR or the social model of sustainability, and irrespective of historical beginnings, named lines of development, core tasks, and knowledge objectives, in practice the management of media companies oriented toward the common good always has to translate the dominance of material objectives into concrete objectives that are as clearly structured as possible, in order not only to justify their existence, but also to enable economic thinking and action by adding the secondary condition of the efficiency principle.

But similarly significant for implementation are the cultural framework conditions and on a social market economy based on a set of values. In order to ensure that the concepts mentioned above do not mean empty labels but real change when they

[16]Neuberger (2012), p. 87ff.

[17]Reising (2019), p. 212ff.

are implemented, a key factor is an extent to which the culture of the organization and the cultural acceptance of change by the public are pronounced and allow for change. In a structured approach to sustainability management with strategies and goals of the organization, it is important not only to consider the tasks, the steering, the focal points, or the analyzed fields of action, etc., but also to know the organizational culture and the expectations of the many intrinsically motivated employees, so that the change of the media house can be supported sustainably in each business area and the sustainability management is not built on sand. An implementation that is anchored and lived in the corporate culture promotes success in an environment that often does not demand personal responsibility or does not enable it, but rather responsibilities are defined, and presupposes personal responsibility of the employees or their empowerment, which is an organizational and individual learning process.

In this context, the management of CSR or sustainability serves to ensure that public media companies align their actions in social responsibility with the Sustainable Development Goals of the 2030 Agenda and the 17 Sustainable Development Goals (SDGs) of the United Nations.[18] Therefore, in the strategy and the goals set by the organization, the social and ecological aspects of sustainable profitability are placed on an equal footing with economic decision-making criteria and, in the spirit of sustainability, are intended to advance these in a balanced or continuous manner. In this sense, corporate accountability serves, among other things, to create binding instruments in the orientation of procurement policy when awarding contracts (mandatory criteria), with which companies are obliged to respect internationally recognized social and ecological norms and standards in particular. From this point of view, corporate accountability in Germany thus goes beyond the concept of corporate social responsibility, which is voluntary for non-listed companies, to promote sustainability, as it creates a legally binding obligation. Since it is easy to describe, but not easy to implement, the holistic and cross-media "model" of the "Gemeinwohl-Ökonomie"[19] (common good economy) or the common good matrix,[20] for example, can provide suggestions and assistance in reflecting on one's own economic activity under the guiding star of common good orientation for practical implementation or approximation. The matrix, which comprises a total of 20 thematic fields and is the prerequisite for the (voluntary) preparation of a common good balance sheet,[21] not only provides a recording system, but can also be used at the formal target level to "measure" the common good-related success by means of a point system of corporate responsibility with the help of the indicators developed in the thematic fields when used periodically.[22]

[18] Mohr and Schiller (2021), p. 3ff.

[19] Felber (2017), p. 205.

[20] Gemeinwohl-Ökonomie Deutschland (2021).

[21] Viest (2017), p. 151ff.

[22] Ulrich (2019), p. 312.

Integrated into an overall management system that links performance and impact targets with financial targets, the core task of management is to develop and introduce concepts and instruments for improving the social, ecological, and economic dimensions in order to safeguard the interests of society and the environment, not only in terms of the media public value of the NPO, but also in this respect. In addition to the identification and implementation of projects and their evaluation based on objectives that are based on the social model of sustainability, internal communication with permanent and freelance employees, especially the demanding generations Y and Z, is also part of this process. In addition to internal communication, it is necessary as a social actor to provide transparent information on the development of the state broadcasting corporation to citizens and society in the broadcasting area (corporate citizenship) within the framework of external communication on the social transformation process of sustainability, which also expresses the cooperation and common objectives in the association of public service media companies.

## 6 Operationalization of Sustainability and CSR in ARD

As already documented in the first sustainability report[23] published in November 2020 by the ARD as a federal media association of the nine German state broadcasting corporations and the German foreign broadcaster Deutsche Welle, they pursue a uniform concept of sustainability. A concept which, on the one hand, is intended to create a balance between the equally important, yet dynamic dimensions of ecology, social responsibility, and economy and, on the other hand, encompasses sector-specific elements of the core business of journalism (e.g., independence, diversity, quality, and regionality) derived from the value-oriented public value and operationalizes them in the fields of action[24] through a catalogue of measures.

In order to be able to ensure the implementation and criteria-based evaluation of the concepts in the core business and in the value chain of the public broadcasters, the integration of sector-specific elements was absolutely necessary, as the catalogue of criteria[25] of the German Sustainability Code (DNK), which was drawn up in 2011 and transposed into national law in 2017 as part of the CSR reporting obligation, is not very specific to the media.

[23] ARD (2020), p. 10f

[24] ARD (2020), p. 12ff

[25] Mohr and Schiller (2021), p. 4.

## 7 Fields of Action of CSR and Sustainable Management in Operational Practice

The focal points and results communicated in the above-mentioned reporting for the entire media network were to be implemented at the institution-specific level through projects or in-depth (individual) measures. Regardless of the time frame, the balance between socially just, ecologically sustainable and economically feasible must always be sounded out.

In operational practice, minimum standards have been defined in the ARD media network in the ecological *dimension of ecology*, in production for example, also in the core business of the organization, which, such as the involvement of the Green Consultant and the balancing of accounts, have been fixed as mandatory criteria. In the case of sustainable media production, for example, a standardized final report must be drawn up after each commissioned Film, television, online and video-on-demand production within the framework of the above-mentioned balancing process, which deals with a large number of individual measures in order to ensure the most resource-efficient, low-$CO_2$ production possible (for example, the use of certified green electricity, the use of rechargeable batteries, the use of LED lighting only, at least 50 percent of overnight stays in accommodation with designated environmental measures, at least 50 percent of the food used from regional sources). The Final report is thus intended to account for the fulfilment of the ecological requirements and at the same time serves to calculate the actual $CO_2$ emissions of the production.

However, in addition to the outlined goals and individual measures in the fields of action of sustainable media production (Green Production) and sustainable event management (Green Events), environmental and energy management are essential fields of action within the framework of sustainable operational management, outside of production operations. These include sustainable canteen supply (e.g., reduction of disposable articles, products from fair trade, and sustainable cultivation), central energy management (e.g., use of renewable energies, modernization of ventilation systems, and use of LED lighting), and biodiversity on the premises (e.g., by planting wildflower meadows, redesigning green areas or through tree sponsorships), but also the supply of green electricity in the ARD network from 2022.

In addition to the employee rights guaranteed by law, the *social dimension* is regulated in particular by collective bargaining regulations of the respective broadcasting organizations, fundamental aspects such as equal opportunities or equal pay for women and men. Furthermore, within the framework of sustainability management, a wide range of further-reaching measures are also implemented in sustainable human resources management, such as supporting employees in caring for their relatives or objectives such as health promotion through ergonomic workplaces or the introduction of a company health management system.

However, in addition to the outlined goals and individual measures in the fields of action of sustainable media production (Green Production) and sustainable event management (Green Events), environmental and energy management are essential fields of action within the framework of sustainable operational management, outside of production operations. These include sustainable canteen supply (e.g., reduction of disposable articles, products from fair trade, and sustainable cultivation), central energy management (e.g., use of renewable energies, modernization of ventilation systems, and use of LED lighting), and biodiversity on the premises (e.g., by planting wildflower meadows, redesigning green areas, or through tree sponsorships), but also the supply of green electricity in the ARD network from 2022.

In addition to the employee rights guaranteed by law, the *social dimension* is regulated in particular by collective bargaining regulations of the respective broadcasting organizations, fundamental aspects such as equal opportunities or equal pay for women and men. Furthermore, within the framework of sustainability management, a wide range of further-reaching measures are also implemented in sustainable human resources management, such as supporting employees in caring for their relatives or objectives such as health promotion through ergonomic workplaces or the introduction of a company health management system.

In the *economic dimension* and its fields of action, such as sustainable printed products, the goal of reducing the use of paper is practiced, for example, through paperless booking processes or the use of document management. The holistic understanding of the CSR concept and sustainability is also implemented in sustainable investments, e.g., by shifting investment funds to correspondingly sustainable business models in financial investments, taking into account ecological, social, and governance factors.

## 8 Conclusion

As a public (cultural) good, public broadcasting in Germany, with its servant character for citizens and society, is per se committed to the common good through its sovereign task of guaranteeing basic services and through the generation of media public value. However, in addition to the special responsibility of their journalistic products as a medium and factor on the part of the public, these non-profit organizations are now not only measured by the success of their diverse program bouquet and its perception or impact, but also to an increasing extent by their responsibility as an economically active actor. Against this background, the management concepts of corporate social responsibility (CSR) complement the respective organizations on the level of formal objectives and are integrated into the overall social process toward more sustainability. Thus, public media companies are not only expected to think and act multidimensionally on the objective level of journalistic responsibility but must also actively live and document their social responsibility as (media) companies in their everyday operations from within. Early and self-initiated adjustments, and not only organizational ones, are therefore essential for the "fitness" and agility to take this social development into account in order to be able to successfully meet the challenges of a twofold social responsibility in the future. In the sense of a credible social change agent or employer branding, the sustainable objectives within the framework of internal and external communication should not only serve transparency, but also the provision of public services and the legitimization of the public media houses.

## References

Alford, H., & Hughes, O. (2008). Public value pragmatism as the next phase of public management. *American Review of Public Administration, 2.*

ARD. (2020). https://www.ard.de/die-ard/wie-wir-funktionieren/ARD-Nachhaltigkeitsbericht-102 (12. Oktober 2021).

Felber, C. 2017: Gemeinwohl-Ökonomie. Eine demokratische Alternative wächst. 3. Auflage, Wien.

Frey-Vor, G. (2017). Qualitäts- und Erfolgsmessung in einem öffentlich-rechtlichen Medienhaus. *Media Perspektiven, Heft, 2*, 106–119.

Gemeinwohl-Ökonomie Deutschland. (2021). https://web.ecogood.org/de/unsere-arbeit/gemeinwohl-bilanz/gemeinwohl-matrix/ (20 Dezember 2021).

Gläser, M. and B. A. Kühnle. 2020: Handbuch Mediencontrolling.

Gläser, M., Reising, W., & Schöne, J. (2016). Performance measurement crossmedial verwerteter Radioprodukte. *MedienWirtschaft, Heft, 1*, 32–47.

Keuper, F., & Brösel, G. (2005). Zum Effektivitäts-Effizienz-Dilemma des öffentlichen-rechtlichen Rundfunks. *Zeitschrift für öffentliche und gemeinwirtschaftliche Unternehmen, 28*, 1–18.

Meynhardt, T. (2008). Public Value: Oder was heißt Wertschöpfung zum Gemeinwohl? *Der moderne Staat, Heft, 1*, 73–91.

Mohr, I., & Schiller, D. (2021). Nachhaltigkeit und gesellschaftliche Verantwortung im öffentlich-rechtlichen Rundfunk. *Media Perspektiven, Heft, 1*, 2–16.

Neuberger, C. (2012). Der öffentlich-rechtliche Rundfunk im Internet: Ein kommunikationswissenschaftlicher Ausblick. *Public Value. Was soll der öffentlich-rechtliche Rundfunk für die Gesellschaft leisten? Beiträge des Initiativkreises öffentlicher Rundfunk., 5*, 31–38.

Reising, W. (2019). Öffentlich-rechtlicher Rundfunk: Jenseits von Staat und Marktversagen – Entwicklung und Etablierung eines wirkungsorientierten Controllings als Antwort auf den Legitimation- und Erfolgsdruck öffentlicher-rechtlicher Rundfunkanstalten. In: *Zur Ökonomie gemeinwohlorientierter Medien – Massenkommunikation in Deutschland, Österreich und der Schweiz. Reihe Medienstrukturen, 14*, 209–225.

Trommershausen, A. (2011). *Corporate responsibility in Medienunternehmen*. Halem.

Ulrich, P. (2019). Ethik, Politik und Ökonomie des Gemeinwohls – Praktisch-philosophische Aspekte tragfähiger "Gemeinwohl-Ökonomie". *Zeitschrift für Wirtschafts- und Unternehmensethik, 3*, 296–319.

# The Dual CR Responsibility of Media Companies: We Only Create Entertainment, Don't We?

**Katja Bäuerle**

## 1 Why Do We as Television Producers Have a Responsibility for Our Content?

UFA reaches about a quarter of the population—one in four Germans—with its programs every week, and one in five among young adults aged 14–49 years. Across all playout platforms, we reach people from a wide range of socio-cultural and socio-economic backgrounds, some daily or several times a day. Among other things, we tell fictional realities of life, we tell who can be like a hero or heroine, we offer narrative responses and possible solutions to a variety of problems, including in our factual formats or the shows. We offer glimpses into other lives, influencing how they are experienced by someone who may not have had such touches in reality. Viewers possibly experience and see things for the first time with us and we make a reflection possible at all. We have the responsibility to entertain, that is our company goal besides economic success. But our responsibility goes far beyond that, given how many people we reach in different situations in their lives today. In addition, with our formats and the people involved in front of and behind the camera, we have a relevant communicative range beyond the original platform and its reach.

K. Bäuerle (✉)
UFA GmbH, Potsdam, Germany
e-mail: Katja.Baeuerle@ufa.de



F. Weder et al. (eds.), *CSR Communication in the Media*, CSR, Sustainability, Ethics & Governance, https://doi.org/10.1007/978-3-031-18976-0_17

## 2 Communicative Reach Beyond the Original Playout Channel

Viewers follow the formats they like and the actors they love on many platforms, and both the actors in our formats and the program makers behind the camera are popular discussion partners on a wide range of topics. With much of what they say, they contribute to a broad formation of opinion, also shaping opinion, but at least with the potential to influence opinions. In terms of strength, immediacy, and number, the parameters of social impact theory, this would certainly have a high "score," especially because here as a "group" our content and our protagonists, whether fictional or real, whether in front of or behind the camera, are inextricably linked. Everything we say has UFA written all over it, and we have to be aware of that at all times. That means we need a broad consensus in the company about the fact that we have a social responsibility and a common lowest denominator on our stance on various socio-political issues, without curtailing the diversity of opinion. And here we are already at the first very striking intersection of our responsibility as a company and as an employer. Every time we talk about our products in public, we are also talking as an employer or employee. We have a very public role model function as a content producer as well as an employer. And in hardly any other industry is it so easy to draw a comparison between the fiction of our products and the reality of our working environment, and thus our external image as a company. But it is not only in our external image that the credibility of our content also depends on our corporate behavior. Internally, too, we have to live with attitude and responsibility in order to empower creatives to engage critically and innovatively with old and new content and topics and to be an attractive place also for future employees who can provide us with urgently needed new perspectives. So, responsibility for our content brings with it responsibility for our corporate actions, and the two areas are not only highly interdependent, they are actually mutually dependent. And it follows that we have to take on this responsibility in an active and formative way, also in order to secure our success in the long term.

## 3 So How Does UFA Meet This Dual Responsibility?

For a long time, we have consistently but rather selectively and sometimes also rather reactively dealt with various issues, such as the topic of ecological sustainability, which gained momentum primarily through an employee initiative, or the topic of women's and men's images in our formats as a consequence of the results of the study published in 2018 on the topic by the MaLisa Foundation. One step taken by the UFA management toward a more sustainable management of our responsibility was, among other things, the establishment of the position of a corporate responsibility management. By establishing such a management staff position, the topics automatically moved forward into the strategic goals and gained new visibility internally and externally, in terms of content but also in terms of their fundamental relevance for UFA. The major areas in CR management for UFA are diversity, equity and inclusion, ecological sustainability (green shooting), and understanding democracy and shaping political opinion. We have set ourselves the goal of portraying German society as it is, both in front of and behind the camera, and stand for equal opportunities for all marginalized groups. We will reduce our $CO_2$ emissions in a relevant way in order to do our part to ensure that future generations can also lead a good life on this earth. We will defend democracy as a political and social system at all times. We continuously educate ourselves and believe in the necessity of lifelong learning. But a CR position can only manage these topics. They must necessarily be developed, designed, and implemented on a daily basis by a broad cross-section of employees. These beliefs and attitudes must play just as much a role for HR managers who formulate job advertisements as they do for producers who draw a character together with a team of authors. The still fairly new D&I department, for example, works in the form of a Circle with employees from different areas of the company. Both the Diversity Circle and the UFA Green Team have their own budget and can formulate goals and adopt measures with a great deal of autonomy. The self-image of these working groups is not necessarily to be an extended arm of the management but to share the responsibility of moving UFA forward and making it better, to significantly influence our Purpose. Of course, the work is carried out in close coordination with the management, which must have the final say in major strategic decisions and also uses these committees in an advisory capacity. Nevertheless, the trust of the management in this committee work is of elementary importance. Continuous collaboration and feedback loops with colleagues from all areas of the company are at least as important as close coordination with management. Just as UFA's CR work originally arose very much from the work of a committed workforce, which then regularly received reliable management backup, ideally the decisions on attitude are made by many troffen in order to repeatedly arrive at a broadly shared UFA attitude and goals that unite as many employees as possible. In many cases, new ideas come from the various working groups, are then tested, discussed, questioned, and improved across various colleague groups until the result, a measure, a tool, or similar, is presented to the management. The goal must be that everyone feels responsible for our products, for what we produce, and how we produce it. If, at the same time, we empower all our colleagues to question each other again and again, i.e., to remain capable of criticism, we will also have a valid safety system in the event that an individual misses something.

## 4 UFA's Commitment to Social Responsibility Also Leads to Further Changes in our Daily Work and, of Course, Brings with it Numerous Challenges

UFA has definitely gained attention through communication and open reflection on its handling of its own responsibility. We notice a great deal of interest in our topics from trade media but also the daily press. We are quoted by third parties and asked for statements in qualitatively different areas, far beyond the focus of our content and products. We have become an even more attractive potential partner for campaign work of various interest groups and the quality of the panels and talks we are asked for has also changed. This is in no way to say that we used to have inferior requests, just that we are now being asked even more regularly about a wide range of topics, also as employers and companies, for example, regardless of individual program formats. However, our transparent engagement with these topics naturally also results in numerous challenges, because the demands on us are high from all sides. Our employees justifiably demand a great deal of open communication and discussion on numerous topics, which UFA must and does face up to. This is fundamentally important, because it is a learning process for all of us each time, but also requires a lot of time, especially from the management, which must see itself as responsible here and also sees itself as steering this discussion.

In addition, we have to constantly re-evaluate which topics we have to or should comment on and where we should and should not focus our efforts. Does a topic have an actual overlap with our responsibility as media creators and producers, where we are competent and experienced enough to enter into a confrontation if necessary? And where we have already formulated a clear internal stance on social issues, how do we deal with it when employees or other individuals or professional groups associated with us take a different position, for example, on social media? Our internal communication needs have increased extremely. At the same time, we must also be aware that we must remain credible and coherent in our actions at every point. We offer a new, quite large field of activity, internally and externally, which provides space for criticism of a new quality. The danger that priorities change for various reasons is always there, we must be aware of this and, if necessary, stand up to it in order not to lose sight of our goals. We are going down this path as a large and stable company that can afford to devote itself to these issues. But responsibility also arises for us from precisely this privilege. The assumption, or rather acceptance, of our own responsibility as a media creators and as a company stems from an inner conviction on the part of everyone involved. Of course, we remain an economically thinking, motivated, and acting entity. Whether all of these points will have a direct economic impact is difficult to say or quantify today. It is rather unlikely that we will get more commissions because we are aware of our social responsibility, have an attitude and also draw consequences from it. And no viewer will watch an UFA format for these reasons either. But we expect a positive change in terms of employee satisfaction and our attractiveness as an employer, which we can already see in current surveys. And our current and future colleagues are the ones who make our programs better, more exciting, more entertaining, and more innovative, who together take responsibility for our company and the economic success of UFA.

# Interviews

**Franzisca Weder**

## 1 "We Need More Sustainability Leadership"

### *1.1 Interview with Mag. A. Malli, Chief Officer for Environment and Sustainability, ORF*

**Weder:** What is your role and what are the main areas of responsibility, if we look at the Austrian media landscape and the ORF as public broadcaster?

**Malli:** My name is Anita Malli. I am now working for the ORF, after leading the "Mother Earth Initiative" ("Mutter Erde"), which is a joint environmental initiative between the ORF and the largest Austrian environmental and nature conservation organizations. This is still part of my responsibilities because I am the managing director of this association. Now I am also the ORF-sustainability officer and in this function, I am responsible for CSR, i.e., for the sustainability reporting of the company and for all processes around "green production" in film and television. This includes the implementation of new ecological and resource-friendly production methods in the television sector.

**Weder:** What is the role of public service media companies in a changing society, a society in a crisis?

**Malli:** We still believe that media are the fourth power in the state. Of course, the ORF should be first and foremost the public watchdog in a democracy. The rulers, the elected, so to speak, all politicians, but of course also large corporations should be "monitored" in a certain way; a public broadcaster like the ORF like any other public service media corporation needs to put their finger on the sore spots and create a public for it. This means creating publicity for certain issues, for relevant topics. This is, so to speak, the role of the media and of public law, of course. In the case of

F. Weder (✉)
School of Communication and Arts, University of Queensland, Brisbane, QLD, Australia
e-mail: f.weder@uq.edu.au

 
F. Weder et al. (eds.), *CSR Communication in the Media*, CSR, Sustainability, Ethics & Governance, https://doi.org/10.1007/978-3-031-18976-0_18

ORF, for example, we refer to the "ORF Act" (ORF Gesetz), which includes a mandate for information.

**Weder:** What about the "corporate social responsibility" that goes beyond the "ORF act" (ORF Gesetz)?

**Malli:** As a company, the ORF is very keen and active under the umbrella of "corporate social responsibility." This is executed mainly through reporting, but also, for example, through the initiatives that ORF also has in the field of "humanitarian broadcasting."

We perceive the main responsibility of the ORF as being the *social glue* that holds the society together. This will intensify in the coming years. We are in a multiple crisis scenario, we have had the Corona pandemic, we are in the midth of the climate crisis. Nothing that we've seen before. To be the social glue here means to contribute to the fact that society is needs to be held together. That's what we perceive as our social responsibility. Beyond this, we do care about our own impact as company, on the environment, for example. For example, we work towards climate neutrality, where the ORF has also imposed its own self-commitment climate targets.

**Weder**: Can you give me an example of a specific CSR activity, for example, in the environmental dimension? An example of ORF living up to its responsibility?

**Malli:** Well, only last year we adopted voluntary commitments, our own "Climate Pact," so to speak, which goes hand in hand with a very large action plan. Part of it will be aimed at consuming less energy. That's one part. And the second part is that, for example, in the area of pre-production, which also requires a lot of energy. Here, we want to establish and further institutionalize the eco-label for "Green Media Production"; we are currently in the process of establishing this production method in-house. But there are other climate targets until 2030. Nothing happens overnight, but it's the *change in consciousness* that matters—and which always needs to go hand in hand with measures, for example, to fade out of $CO_2$-high emission cars in our TV/broadcasting fleet, but there is a huge package of measures.

**Weder:** The ORF is a company that produces a very special product, a journalistic product. How can responsibility be exercised *within* this product or with this product?

**Malli:** This is a double-edged sword. On the one hand, we have editorial freedom, which is the highest good. That is, in reality, up to each editor to decide whether he or she interprets this responsibility for himself or herself. In reality, as a journalist in a public broadcaster, you see yourself as being responsible to your audience, committed to the audience, and to the ORF law. And here the highest good is objectivity. In this respect, it is not so easy for a company like ORF. Especially in times of great crises and in times of the climate crisis, it is good that this editorial freedom remains.

**Weder:** So, in the context of CSR, are we talking about journalistic responsibility? Or institutional responsibility?

**Malli:** The point is that you can't get around it. Journalism and what it is, or what it should do, is directed by and somehow the definition is reserved for the editors. However, I can already see a certain responsibility of those editors and editorial directors, who encourage journalism to consider the essential role of the climate. But it is ultimately the responsibility that we need managers who also want this—and enforce it.

Above all, however, we need training. We as the ones responsible for sustainability within the ORF can offer the trainings, we can organize them. But the editorial director must want it and to push for his people to go there, his people to have the know-how to inform about climate change and sustainability.

**Weder:** What kind of trainings are these?

**Malli:** For example, basic climate knowledge. We had this one basic training, where our "weatherman" introduced basic information about climate change, weather, and related climate facts. As a journalist, you simply have to have heard—and understood—the essential arguments in the debate about the climate crisis. You must understand and to decode sometimes very scientific information; and to decipher the reservations and the skepticism.

**Weder:** Are there also trainings specifically on sustainability?

**Malli:** No, we actually focus specifically on the climate crisis. But if a journalist needs some background, he will find something. The difficulty here lies in the concept of sustainability, because it is very thin and watered. And then it is all about "green" and "greening" different issues and topics; only very few people out there know that there is social sustainability and not only environmental sustainability. The question is also: What is sustainable in the first place, or on a more general level? And a lot of misunderstandings around the term, a term that is abused by companies quite often. Think of Sustainable Aviation Fuel, for example. Because I add something to kerosine it is not yet sustainable. Flying is not yet sustainable; thus, there is a lot of disinformation around sustainability—not easy to talk about from a journalist's perspective.

**Weder**: Is there any way out of this sustainability dilemma? What can a media company do?

**Malli:** First of all, you need a clear decision to deal with what sustainability actually is. Okay, it's a megatrend. But it is important to keep in mind that it's really about negotiating what would actually be sustainable, what activities and behavior would be sustainable, as individuals and as organization itself.

Until the 80ies, we were always pretty sustainable, since then it has gotten worse and worse. But the main question is: where is it being talked about today? Where do we think about and negotiate what sustainability and being sustainable is? This is where I see the responsibility of media companies in particular. It is sustainable not to overuse the resources and to preserve the regeneration of our ecosystems, even as a company, but it is necessary to explain that, to continuously talk about it.

There is the Paris Treaty, everyone has agreed, all parliaments have ratified. And what does that look like now? Although there is no compulsion, no one talks about it, but like you now, for example, in a political reporting what does the FPÖ's plan for this look like? What is the NEOS' plan for this? What is the SPÖ's plan for this? And what is the plan of the ÖVP and the Greens to achieve these goals and then compare them, for example? If you're a public watchdog (like the ORF should be), you'd have to take a close look. Because it depends on that, whether Homo sapiens has a future.

**Weder:** Is this an Austria-specific problem? Is it an ORF-specific problem or simply because of the size and importance of this company? Or is it really a general problem of media and journalism?

**Malli:** I think this is an international problem. I mean, there are already media outlets that do it well. An example would be that a sports journalist also works as a business journalist when he can report on the transfer of Messi also from a financial and economic perspective, not only from a sports perspective.

Or think about the financial crisis of 2008. The financial crisis was the big issue. It has encompassed all areas of the society. And there were also an insane number of training courses for journalists, so that all journalists, even those who had to deal with economic and financial issues for the first time, knew what was going on. If, for example, a football club suddenly runs out of money, the sports journalist must also know approximately what is happening and what the background is. And that's how it has to be now in the climate crisis! Every journalist needs to be capable to look at other issues with the climate crisis lens! While enforcing and implementing this is, in turn, a management task within the editorial offices, for example, to simply demand and facilitate the training courses I've been talking about.

**Weder:** How would it be if the ORF had this climate lens or "tangent" in every story? Would that not be well compatible with their idea of creating "Public Value"?

**Malli:** Yes, it would be! But many colleagues complain about attempts to exert political influence and are sometimes still seen as activists when they repeatedly raise the issue.

And that's just a big problem—also in other countries, such as the USA or Australia. Whereas in the UK the topic has already arrived, especially at the BBC. The BBC is certainly a pioneer in this respect, from cost models to broadcast formats and the trainings for journalists.

Climate and sustainability reporting as a business is one thing. Reflecting climate change in all topics that are reported and communicated about in our media corporate, that is, something else. This must start from the leadership and then happen at the program level. If every journalist goes into those training and awareness grows bottom-up, then we don't need to talk about great CSR programs anymore. The seriousness or the magnitude of the climate crisis in all ecological and social dimensions, this is, a special task for the corporate management for sure.

**Weder:** Thank you very much for the interview!

# 2 "Sustainability Has Always Been an Intrinsic Value for Us"

## *2.1 Interview with Prof. Dr. Jens Müller, Department of Strategic Corporate Development and Media Policy, ZDF*

Rademacher: Dear Mr. Müller, ZDF recently submitted another sustainability report and issued a declaration of conformity. Can you briefly describe how ZDF approached the topic of sustainability: How did you get to the point where the station is today?

Müller: ZDF was first explicitly confronted with the issue of sustainability in 2004, when the Council for Sustainable Development wrote a letter to the director and also to the managing directors of all other German media companies, TV stations, and so on, asking them to prepare something like a sustainability report. That ended up with me. We then compiled a sustainability report to the best of our knowledge and belief at the time, which would probably no longer meet any reporting standards today.

From 2010 onward, we regularly published sustainability reports that followed our own signature, our own storytelling, quite opulently designed, on three levels: Company, industry, and society. In them, we tried to present what ZDF does beyond its programming as a commitment to the company, the industry, and society. These were always brochures as PDFs of about a hundred pages. And then in 2014, the EU directive followed, in which it was already indicated that sustainability reporting would become mandatory—even if it was not entirely clear which company size would be affected. At ZDF, we preemptively agreed with the management at that time to follow one of the established reporting standards. And since the UN Global Compact was a bit too short for us, but the GRI standard was too extensive, we decided on the German Sustainability Code and have just submitted the third declaration of conformity in July 2022. This declaration of conformity has, so to speak, replaced or supplemented the part relating to the company in our original sustainability reporting.

Rademacher: Is that how you still see it, that this is actually a voluntary report? Is that still how it is handled?

Müller: Yes, this is still a voluntary report, at least according to the lawyers with whom we are discussing this. We know that there are still a lot of things happening in Europe at the moment as far as regulations are concerned: CSRD from 2024, the taxonomy, etc. However, we believe that, despite all the voluntary nature, in a society, in a market in which many follow such reporting standards, an obligation arises because the companies deal with their suppliers or other stakeholders, for example—and perhaps also associate a certain reputational effect with the preparation of the report.

So, we at ZDF believe that it makes sense to present these reports, especially since the internal effect should not be underestimated. Because when you do something systematically, goals automatically become much clearer. For example, we are just starting a mobility analysis of our employees in order to gain transparency for us as to which Scope 3 criteria we already fulfill for the GHG Protocol. In this way, we are trying to further sharpen our sustainability reporting.

Rademacher: If we could please take a step back from the specific reporting topics for a moment: What is the role of media companies in society?

Müller: Well, the media undoubtedly play a very important role in society, in politics, in the development of societies. If I now relate this to my company, originally a public TV company (Public Service Media, PSM), which is of course expanding a bit in the context of media libraries and nonlinear TV consumption, I would say that a contribution-financed broadcaster like ZDF, in particular, must be aware of a special responsibility. And here I would put the issue of sustainability right at the top. You can put that in concrete terms in many places, such as the status of suppliers, the treatment of employees.

The second thing, which is of course the main impetus for a media company, is ultimately something that I would perhaps call public value in the context of public television (PSM), or in economic terms positive external effects, because of course media with their content of a journalistic nature have a very special responsibility, especially in times of social media, of filter bubbles, of fake news.

And that would also be my only slight criticism of such reporting standards: that they do not even consider the level of the products. After all, a TV format produced in a green way would always do better in reporting than one produced in a less green way. But in my opinion, the content level is much more decisive. Then, to use a perhaps somewhat hackneyed example, the mafia boss who collects his protection money and drives up in a Prius is probably rated well in a reporting standard because it is $CO_2$ neutral. And in my opinion, the core of a media company, especially one under public law, is an added social value such as the contribution to integration, to democratic decision-making. That may sound a little exaggerated, but I think it's very topical at the moment.

Rademacher: But that means you would still give the concept of public value a strong meaning today? The term is rarely used anymore.

Müller: Yes, I would actually use the term public value. You are right, of course, because many of these terms now seem to be used in an inflationary manner in our sustainability context. So, if we talk about purpose and impact, we would both just shake our heads and say: Whether it is a food company with a high use of fat, perhaps palm oil and sugar and very little cocoa, or an operator of coal-fired power plants, they may also write such terms on their flags. Which is fine, if sometimes it wasn't just in public relations or even greenwashing. But for me, public value actually refers to a positive external effect—in economic terms—the core of what you have to do when you work as a contribution-financed broadcaster and not as a marketer of advertising time.

Of course, the business model of traditional media companies is fine, generating attention with their business model and ultimately converting contacts into advertising revenue. But that also entails the risk of perhaps appearing less investigative, less critical, and also less inclusive, namely because it is more oriented to the reaction of the audience and their filter bubbles.

Rademacher: You had already talked about the EU requirements from 2014, which all came under corporate social responsibility. In your view, to what extent is that a parallel concept to sustainability? And to what extent does the term also apply to media companies?

Müller: You are absolutely right. The terminology in our context—corporate social responsibility, corporate responsibility, sustainability, corporate citizenship, and so on—is not entirely free of overlaps, and there is also a great deal of redundancy. When I tried to explain this to students, I always tried to distinguish between the goal of the whole: Sustainability according to the classic Brundtland's definition of sustainability: "Leave your children a world as you found it, with all the corresponding opportunities." So sustainability as a goal and CSR as a strategy. Now the term CSR, so to speak, if you look at it in terms of its usage, was primarily set with the S for social. And balancing the triad between economy, ecology, and social issues was not always easy (triple bottom line).

I found it quite appropriate that the term sustainability was used to focus more on climate and environmental issues, such as climate neutrality, climate change, $CO_2$, and so on.

And if we now look again at the CSR Directive of 2014, the term diversity played a major role. I think this has shifted a bit out of focus, but in my opinion, it is already included in Brundtland's original definition. However, the original definition of CSR in the European Commission, as I understood it, went beyond what was required by law. However, it is precisely this aspect of voluntariness that seems to play less and less of a role today.

Rademacher: What all belongs in a definition of CSR today, as you would represent it for ZDF?

Müller: Well, I would say that social and ecological criteria really do play an important role, and I think that the 20 points of the German Sustainability Code are very useful. Of course, they provide a framework that is by no means sufficient to describe and reflect everything that happens in detail in facets beyond that in a media company. So, when I look at diversity according to reporting standards, it's a typical "bean counting": the proportion of men and women with and without an immigrant background by age. And we often see in these reporting systems this tendency to "bean counting," sometimes just out of embarrassment: what I can count, I count. What I can measure, I measure. But it is very difficult to measure the actual dimensions of diversity, the lived dimensions. It's the same with green production: I can measure the $CO_2$ emissions of an editor's flight from Hamburg to Madrid quite well, but I'm perhaps less able to measure the actual impact of the contribution or content she made.

If, for example, it calls for sustainable action, then the effect is much, much greater. This means that, beyond the reporting standards, we actually have to have our own storytelling that is adapted to the industry and the company. At ZDF, we solve this by communicating our own commitment to the outside world in these different social, societal and ecological criteria, which we then describe in terms of content.

Rademacher: The website also shows the station's commitment to content, and you get the impression that the editorial work on climate issues, including the topics you just mentioned, such as cultural diversity and so on, which also occur in the program itself. And that is then flanked by statements from the management and others that are collected there. All in all, the impression is that this content-based approach to the topic of sustainability is actually part of the station's sustainability strategy. Is that the case? Or is the editorial focus on sustainability a kind of reflection of reality, or even part of the program's mission as you understand it?

Müller: To be honest, I would turn it around: For me, the sustainability strategy and reporting are part of it, but the core mission, which is based on the constitution and secured by numerous rulings of the Federal Constitutional Court, this core mission is to educate, inform, entertain, reproduce content and so on that reflects social reality. This is a mission that in principle reflects what we call a public good in other contexts of the state's mission. A public good would otherwise probably not be provided by the market. And this provision of an important public good, with corresponding positive external effects, is primarily the program. And fulfilling this mission is not just a matter of working with stakeholder groups to create a suitable materiality analysis. In some companies, I find that these materiality analyses are a cover to sell the prioritization and that what ends up in the top right-hand corner, in companies that perhaps produce unhealthy food, suddenly looks really, really great and green.

In our case, we are in a permanent stakeholder dialog because of the close ties with our committees—the TV board and the board of directors—and because many social groups are represented there. And that's not always easy.

I have to say that I wouldn't want to miss the differentiated and very diverse composition of our committees, because ultimately, they also represent a legitimation context for the contribution privilege. Of course, not all social groups are willing to pay for such a public good. But that is generally the case with public goods, where everyone benefits from something that not everyone would be willing to pay for.

That's why it was very important to me to include this long quote from the ruling of the First Senate of the Federal Constitutional Court on the First State Media Treaty at the very front of our last sustainability report. The Federal Constitutional Court argued very well here what makes us so special in an age of filter bubbles and fake news.

Rademacher: In your view, is there an explicit mandate for ZDF to make an active contribution to making climate justice or climate neutrality a central value in society?

Müller: I think so. I believe that public broadcasting first of all reflects the environmental factors. If you formulate this somewhat simply in terms of business management in a SWOT analysis, then in the meantime, the fields that you have for the weaknesses and threats are probably completely overfilled. These quadrants would probably have to be much larger than the others, especially with regard to climate change. And I think ZDF's program reflects this risk dimension quite well. And tries to convey how you can deal with it.

First, there is the global level of reporting. Then there's an educational level, giving advice on what you can do, what you should perhaps not do. There's a whole section, in the documentaries, that deals only with climate change, with climate neutrality. There's a very strong exploration of "living sustainably." And it's all done not in a detached way, but in a critical and accessible way. We do this in a style that is probably also quite suitable for positively influencing behavior in the long term. As a TV station, we can only influence this in a very concomitant way, but the aim is to take up and reflect this aspect of climate change and its intensified progress, and to ensure that the world becomes, if not better, then at least a little less bad from this point of view.

Rademacher: If you agree with that, then we are not far away from constructive journalism or solution journalism. Does that then mean: the public education institution of public broadcasting?

Müller: The term constructive journalism has now been clearly defined. However, if I interpret it semantically, I would say that journalism should always be constructive. I already know what you mean: namely, to be active in journalism with appropriate effects. I believe that public television and ZDF, in particular, have been constructive in their coverage of this issue. However, you also have to bear in mind that we naturally have something like program autonomy. The journalists in the editor-in-chief, and also the employees, editors in the program management—these are our two program-forming areas—first of all act autonomously, so that a director or a sustainability officer cannot now tell them: "Please now take up a certain topic in the context of sustainability." And the stakeholders, who would of course like to exert their influence, have no access here either. In my opinion, the journalistic freedom of our employees with regard to such topics results in constructive journalistic implementation.

However, this cannot be embedded in a sustainability strategy at the push of a button, as it were, along the lines of: Do this for a change. But people have always been intrinsically motivated to do so. With our "Planet e" format, we had the first programs ever here, starting in 2011, that explicitly dealt with the topic of the environment with its own broadcast slot. At the time, that didn't exist anywhere else. And the topic of sustainability, whether it's "Lesch's Cosmos" or an advisory format on radio or YouTube, is per se a concern for journalists working in public broadcasting. Because, of course, a critical view of social, economic, and even climate developments is part and parcel of journalistic engagement with the world.

Rademacher: How do you assess the current trends around sustainability? Every product advertises with it today, but at the same time people know less and less what is actually meant in concrete terms. What does this development mean for you?

Müller: It's true that the topic has gained a lot of momentum in the last 20 years, if we think of the SDGs, if we think of the various reporting standards. And yes: in the meantime, there has even been a kind of overdose, because of course everyone and everything adorns themselves with the label sustainable, even in a completely different context. So that one is almost inclined to say goodbye to this term because it is used so inflationarily. Similar to Purpose and Impact, because they have quickly become part of this rapidly expanding sustainability industry with management consultancies, reporting agencies and certifiers. This has become a business model and a profitable business, and when one is involved in this business, one often forgets that doing good is first of all the core of it, and talking about it well is the second point, and only then systematically "reporting" the third. These are not mutually exclusive, but in the face of all the labeling and marketing and PR, one should not lose sight of the fact that it is perhaps also more important in the personal sphere to separate waste in the first place, before I tell others that they should do so—or even get a stamp confirming that I am the most waste-conscious separator in my rental unit.

Rademacher: Thank you very much, then we'll come to the end: Do we actually need an eighteenth sustainability goal that relates to media? What do you say?

Müller: We don't need a new sustainability goal. But what we do need, however, is what GRI used to have, namely industry-specific standards. There used to be corresponding media standards in the GRI 3 context, but they no longer exist. The DNK, like the GRI or the UN Global Compact, are more like a lawnmower that goes over everything at once.

A standard would have to be adapted to the specific industry. Now every industry will claim that they are very special. But I believe that the media are indeed special in many respects: We don't even have to talk about the two markets, the public goods problem, information asymmetries, and much more. That's why I'm convinced that it would be very expedient to work together in an industry alliance to supplement this reporting standard, in this case the Sustainability Code, at least at the German level, because in the international frameworks such as the SDGs, for example, this can certainly be applied to the media.

But it's easier for an organization like the German Sustainability Council to say: Come on, I'll offer you something. And it's easier for companies to say: Yes, stamp it. "DNK approved!" Declaration of conformity is there. In this respect, yes, I agree with you that we need a media-specific sustainability standard, as we are seeing again right now. The Ukraine conflict and the accompanying war for information and content show this very clearly. Our media are of essential importance worldwide for our social and global development. And we cannot afford to use standards that apply to everyone.

The interview was conducted in mid-July 2022.

The manufacturer's authorised representative in the EU is Springer Nature Customer Service Centre GmbH, Europaplatz 3, 69115 Heidelberg, Germany. If you have any concerns regarding our products, please contact ProductSafety@springernature.com

Printed and bound by CPI Group (UK) Ltd, Croydon, CR0 4YY
28/09/2026
02224071-0008